JACOB'S LADDER

ALSO BY HENRY GEE

Deep Time: Cladistics, The Revolution in Evolution

JACOB'S LADDER

The History of the Human Genome

Henry Gee

W. W. Norton & Company New York • London

Manufacturing by Courier Westford

ISBN 0-393-05083-1

W. W. Norton & Company, Inc.
500 Fifth Avenue, New York, N.Y. 10110
www.wwnorton.com

W. W. Norton & Company Ltd.
Castle House, 75/76 Wells Street, London W1T 3QT

1 2 3 4 5 6 7 8 9 0

And he dreamed, and behold a ladder set up on the earth, and the top of it reached to heaven: and behold the angels of God ascending and descending on it. And, behold, the Lord stood above it, and said, I *am* the Lord God of Abraham thy father, and the God of Isaac: the land whereon thou liest, to thee will I give it, and to thy seed; And thy seed shall be as the dust of the earth; and thou shalt spread abroad to the west, and to the east, and to the north, and to the south: and in thee and in thy seed shall all the families of the earth be blessed.

Genesis 28: 12–14

Wherein I spake of most disastrous chances,
Of moving accidents by flood and field,
Of hair-breadth 'scapes i' the imminent deadly breach,
Of being taken by the insolent foe
And sold to slavery, of my redemption thence
And portance in my travels' history;
Wherein of antres vast and deserts idle,
Rough quarries, rocks, and hills whose heads touch heaven,
It was my hint to speak — such was the process —
And of the Cannibals that each other eat,
The Anthropophagi, and men whose heads
Do grow beneath their shoulders.

Shakespeare, Othello, I, iii, 134–45

Contents

Preface

On 12 February 2001, an international team of scientists announced that they had substantially deciphered the human genome – the genetic instructions for creating and maintaining a human being, and our evolutionary birthright. The event was reported in the *New York Times* as an achievement representing a pinnacle of human self-knowledge. The effort to sequence the human genome has been compared with the Apollo programme to put astronauts on the Moon, so the scientists were surely entitled to a certain amount of self-congratulation.

To describe the sequencing of the genome as a technical feat – like sending astronauts to the Moon, or even crossing the Himalayas on a unicycle – is to miss the point. To be sure, it is fun to learn that if each one of the three billion DNA bases that make up the genome were magnified to the size of a letter on this page, the genome itself would stretch across the continental United States and back again. But such facts stupefy rather than edify. The genome is important not because of what it is made of, but because of what it does: it is the agency that creates and maintains an organism, urging an endless variety of exquisite life forms from formless eggs. As such, it transcends

ix

the particularities of its substance and becomes a motif that has been central to biological thought since antiquity, making the achievement of 2001 all the more profound and exciting.

If any discovery represents a pinnacle of human self-knowledge, it does so only by virtue of the support of the mountain on which that pinnacle is raised. For thousands of years, people have wondered about the identity of the mysterious and marvellous entity that makes babies, shapes our evolutionary history and populates the world with living things of almost unimaginable variety – in short, the thing that teases form from the void. The history of biology can be retold as the story of the search for this agency, the genome. My aim in this book is to tell that story. Or rather, three stories.

The first, and the most direct, is an account of the intricate development of a human embryo from an egg. A pea-sized embryo recognizable as human develops from a fertilized egg in just four weeks, often before the mother even realizes she is pregnant. In this first strand can be seen the initial impetus for this book, which grew out of a desire to express the wonder that every new parent feels on confronting birth, an event which is both intimate and timeless.

The development of a human embryo is at the same time an expression of unique individuality and universal heritage. At this point the first story gives way to the second: how the course of individual development mirrors the history of human evolution itself, back to the dawn of life. Put another way, the human genome directs the development of every single embryo, but is itself a product of evolution and a reservoir of evolutionary memory.

The third story – and the one around which the book as a whole has been constructed – is the tale of discovery, of how people over many centuries have come to understand the development of individuals in terms of the variation, diversity and evolution of species.

As I researched the book, I found to my surprise that the story can be traced back to antiquity and extends more or less seamlessly down to the present. From a fully historical perspective, modern scientific research shows clear evidence of its ancient roots. For example, every scientist takes for granted that the genome in any particular individual, while unique to that individual, is at the same time the vehicle for our heritage. This is not a modern view, but stems directly from a theory which is now all but forgotten – the theory of 'preformationism'. This idea, which formed the mainstream of biological thought in the late seventeenth and for much of the eighteenth century, holds that the germ of each individual is not made anew with each conception, but was created in all its essentials at the beginning of time. In other words, conception does not start a new life from scratch, but simply activates a programme that was already in existence, and which has existed since the beginning of time. The modern idea of the genome as the eternal encapsulation of the instructions to produce a human being owes much to preformationism. Watson and Crick's classic paper from 1953 on the structure of DNA, containing the now famous line 'It has not escaped our notice that the specific pairing we have postulated immediately suggests a possible copying mechanism for the genetic material',[1] is pure preformationism.

To take another example: all modern evolutionary biologists are conscious of a link between the events in the development of any particular embryo and the shape of the history of the species to which that embryo belongs, and indeed the history of all life. Scientists now take for granted that there is a deep

[1] Watson, J. D. and Crick, F. H. C., 'A structure for deoxyribose nucleic acids', *Nature*, vol. 171 (1953), pp. 737–8.

connection between embryology and evolution; and that the shapes of living organisms are not random, but display very clear patterns which are indicative of a shared evolutionary relationship. This insight emerged during the era of romanticism at the turn of the nineteenth century, and found scientific expression in a movement known as 'nature-philosophy', which saw human development as an expression – indeed a culmination – of a universal tendency towards perfection: the 'microcosm' that measured the 'macrocosm'. Nature-philosophy was very much associated with the thought of the great German savant Goethe, and survives today in its purest form in 'anthroposophy' – the view of life developed by the Austrian philosopher Rudolph Steiner. It also turns up, more or less disguised, in various 'alternative' points of view, from homeopathy to ecological activism. Given this heritage, many scientists would be intrigued, even horrified, to learn (if they did not already know) that modern developmental biology grew directly from the work of nineteenth-century German embryologists who had been schooled in the prevailing atmosphere of nature-philosophy. Indeed, after long dormancy, it is now almost possible for a serious biologist to whisper the name of Goethe in a lecture without raising sniggers at the back.

A slightly more recent and telling example of ancient roots traceable in modern thinking concerns the founder of the science of genetics, William Bateson. He realized, in the 1890s, that the nature of genetic variation was completely unknown – a serious problem, given that Darwin's theory of natural selection had been based on variation. This, Bateson saw, was the source of the general dissatisfaction with Darwinism being shown in the last decades of the nineteenth century. His solution was to produce a catalogue of every case of biological variation he could find, in the hope of discerning general laws. The result, *Materials for the Study of Variation* (1894), was fundamental, and in more ways than one. As well as trying to address variation from first principles, *Materials* harked back to an archaic tradition in scientific thought, that of the 'bestiary' – the medieval catalogue of natural freaks and monsters, a genre which early scientists such as Francis Bacon recommended as a useful exercise in understanding the extent of natural variation. But *Materials* is far more than a dry catalogue – it is a work wrought in passion by a scientist determined to uncover the roots of variation and inheritance. Less than a decade after *Materials* was published, Bateson coined the name for a new science – genetics. Through him, therefore, the modern science of the genome has deep connections with the earliest stirrings of biological thought, if not the fumes and smoke of alchemy.

Why, then, should I have been surprised that our modern understanding of the genome has such clearly visible if ancient roots? After all, I did not have to consult ancient grimoires, scrawled in forgotten dialects, locked in the dusty libraries of remote and haunted castles: my investigations led me no further than a small number of reasonably well-known and available texts that can be read with ease while strap-hanging on the London Underground.

I believe that our view of the long history of biology has been clouded and distorted by the titanic presence of Charles Darwin and his book *On the Origin of Species by Means of Natural Selection, Or the Preservation of Favoured Races in the Struggle for Life*, published in 1859. No one who considers themselves open to the facts (and I presume to include myself here) can doubt the pre-eminence of this book in the history of biological thought, and the explanatory power of the theory of evolution that has grown from Darwin's simple mechanism of natural selection. What is less appreciated, I believe, is the extent to which Darwin and his theories were very much products of their time, and that the reputation of both have waxed and waned since 1859. The contemporary Darwin industry has bred a strain of popular science from which it would be easy to conclude that nobody knew anything about anything until Darwin arrived in 1859, as if on a fiery chariot from heaven, and gave the world his graven tablets – after which the scales fell from the eyes of all, and nothing remained to be discovered. I sometimes wonder whether Darwin – a practical man yet given to crippling insecurities, forever worried about the health of his family and his share portfolio as much as matters of science – would have recognized himself in the grotesque monster that is his contemporary hagiography.

This somewhat cartoonish view of history engenders a similarly simplistic view of the history of science more generally, in which we are seen as progressing steadily, as if on a unified front from the past into the future, forever shining brighter lights of discovery into an ever-shrinking puddle of ignorance. As a consequence, if we hear anything at all of biology before Darwin, it is brought up only to be belittled. If the nature-philosophers are depicted as hopeless romantics, the preformationists will be seen as periwigged buffoons who drew little men in the heads of spermatozoa and believed it truth; and the contributions of the alchemists, self-locked in crepuscular dungeons of horror, are to be viewed as entirely undeserving of serious consideration. One of my tasks in this book is to set Darwin in context and show how these earlier sources influenced both Darwin and later scientists, down to our own age. I feel very strongly that this rehabilitation is not only desirable but necessary, as we shall require the fullest and most dispassionate appreciation of the entire historical perspective of biology if we are to face the coming decades, in which our knowledge of the human genome might be applied to alter the Earth – and human beings – beyond recognition.

That time has not yet come. It is evident, however, that the standard of debate surrounding related issues of our own day –

abortion, *in vitro* fertilization, the genetic modification of crops, and so on – is hardly adequate to address even these problems, and yet the prospect of the modification of genomes makes those concerns trivial indeed. The time *has* arrived when we must address – seriously – the relationship between our biology and our humanity. But before we can do that we must understand how we got to where we are now, and it is to this end that I offer this modest contribution.

The genesis of this book was as involved as that of any living creature. It started some years ago with a conversation I had with my agent, Jill Grinberg, in the lobby of the Kitano Hotel on Fifth Avenue, New York City. I thank Jill for her continuing support; I should also like to thank Carl Zimmer for sharing the earliest pangs of this book when I lodged at his apartment in Queens. I thank Peter Robinson and Sam Copeland at Curtis Brown, Christopher Potter, Leo Hollis, Catherine Blyth and Sophie King at Fourth Estate, and Angela Von Der Lippe at Norton, for their guidance and encouragement. I thank the librarian, staff and fellowship of the Linnean Society of London for housing me for part of the time I was sketching the draft, and for finding a number of important sources I would certainly otherwise have missed.

I would never have got anywhere without the indulgence of my colleagues at *Nature*, especially Christopher Surridge and Rory Howlett, who kindly commented on various parts of the draft; and Michael Kenward of the Association of British Science Writers, without whom I would not have discovered Clara Pinto-Correia's book *The Ovary of Eve* until it was too late. Joanne Webber handled the administration with her customary elegance.

A number of others were kind enough to read and comment on earlier versions of this book, either in part or as a whole, at various stages of its gestation. They were Wallace Arthur, Ted

Chiang, Jack Cohen, Craig Davidson, David and Shirley Forbes and family, Walter Gratzer, David and Fiona Hulbert and family, and Charles Middleburgh. My wife Penny hefted the draft across London from east to west and back at unsocial hours. I am grateful to Philip Ball for suggestions and materials related to Paracelsus and Goethe; and to Paul Carline for being an articulate exponent of an entirely different view of life from my own, and whose opinions resulted in many reconsiderations and some radical transformations. I thank John Woodruff for an expert edit. I owe a particular debt to Tony Kerstein, who, in his role as The Man on the Ilford Omnibus, read and commented on almost every draft as it emerged. Tony, the drinks are on me.

Not all these kind people enjoyed the book, and any errors and opinions are my own unless noted otherwise.

In addition, I thank the following for permission to use copyright material: Francis Crick, Macmillan Magazines Ltd, North Atlantic Publications, Oxford University Press, Princeton University Press, Random House UK, The University of Chicago Press and James D. Watson.

Needless to say, all this would have been impossible without the unstinting support of my family – Penny, Phoebe, Rachel, Marmite and Fred. The Cranley is gone, but not forgotten.

Henry Gee
Ilford, September 2003

PART ONE

1

Birth

A little girl is about to be born. Just 200 days ago she was a single cell, a mote quickened into new life, yet made from the salvage of the inaccessibly old; a new combination of genes, passed down like heirlooms from the first men and women; the first fishes that ever crawled from the sea, the first things that were ever alive. Her genes are at the same time a treasury which belongs to her alone, and the heritage that belongs to all of us.

In 200 days she has grown from the edge of invisibility until she is bursting to get out. Her growth has been more than mere inflation. The single cell of a few months earlier has divided and multiplied to become trillions strong, with each cell in its own place in relation to all the others. There is a direction to her organization almost too intricate to follow: the child I shall soon see in all her wholeness is ordered on every level. She will have organs; a heart, arms and legs, brain and skin. Her organs will, in turn, be made of tissues, carefully assembled – ranks of muscles; a frame of cartilage, soon to become bone; neurones wiring themselves together.

Before the microscope was invented, organs and tissues were

thought to be the indivisible atoms of life. Organisms without organs or tissues were not reckoned to be truly alive. Until the microscope revealed it, nobody suspected that organs and tissues had any kind of constituent structure, that the intricate order of a newborn baby goes beyond the limits of human sight – down to the submicroscopic communities of cells and chromosomes, to the genes themselves. My daughter, yet to emerge, will be not only an organism, unitary and independent in her own right, but a vast, tiered community, a hierarchy of systems, built on trillions of interlocking cells and thousands of genes, independent yet interdependent. All this from one egg cell and one sperm cell, in just 200 days.

From a single fertilized egg will have emerged a human, a unique individual with an identity and a name – and yet recognizably the same as the billions of humans already in existence. How can it be that a person can be made in so short a time and with such intricacy? How can a person be made to be unique, yet clearly a member of the human species?

A little girl is about to be born. All of a sudden, she is here. Her umbilical cord is cut and tied. After moments which pass in viscid slowness, she cries. A nurse calms her, weighs and inspects her. She is fine, she is *OK* – *all* of her – and the first step on her journey through life is complete. In many ways it is the most important and the most hazardous, for it is in the first few weeks of life that the genome does the work of a lifetime. In the first four weeks after conception, the genome takes a formless speck and shapes it into what is recognizable as a human being. No other comparable period of human life is as significant, as defining or as busy. The remarkable thing is that the genome can act both so quickly and so reliably: in the time it has taken you to read this paragraph, another 200 babies will have been born. Each will be slightly different from the next, but they will all be human, similar enough to share the joys and pains of life. The genome performs this delicate act of construction, shaping form from the formless, with breathtaking speed and — to judge from the crowds of human beings already on the planet — effortless reliability.

My daughter's voyage began thirty-seven weeks before her birth, when one of many thousands of sperm met a single, receptive egg shed by one of the two ovaries of the mother-to-be.[1] The sperm penetrates the egg, disappearing into its gelatinous confines. The egg responds by expanding outwards, offering a distended envelope as an obstacle to any other approaching sperm. The successful sperm is like millions of others, a package of genetic material propelled by a lashing tail, distinguished only by its luck in having got there first. But in victory lies annihilation. The tail and all the other parts of the sperm are dismembered by the egg; the genetic material, packed into the head, sinks downwards to the nucleus of the bloated egg and merges with it.

A fertilized egg is called a zygote. Its genome is unique — the germ of a new individual — but not new, being made of equal contributions from each parent. Within twenty-four hours, the zygote divides into two identical cells, lying within a common fertilization membrane. Just four days after conception, the embryo has become a ball of thirty-two cells straining at the confines of this same fertilization membrane. At this stage the embryo looks like a berry — in fact it is known as a morula, after the Latin for mulberry. After another day, a small pool of fluid starts to form in the middle of the morula, pressing the cramped cells up against the inside of the membrane. When this happens, the morula crosses the line to become a new stage, the blastocyst.

As these events unfold within the fertilization membrane, the germ of new life floats down the fallopian tube, from the site of fertilization and into the uterus. This is when the embryo performs its first act of defiance: the blastocyst bores a hole through the fertilization membrane and emerges, rather like a butterfly from a chrysalis, to become a free agent. This freedom is temporary, for the naked blastocyst immediately burrows its way into the spongy lining of the uterus, an event known as implantation.

The second week of gestation begins with some internal housekeeping as a knot of cells on one shore of the blastocyst cavity – the pool of fluid within the blastocyst – begins to organize itself into discrete layers. A second fluid-filled cavity forms between two of these layers and the blastocyst wall. This new, second cavity will eventually become the so-called amniotic cavity, whose rupture, months later, signals impending birth. The result of this process is the formation of the germinal disc – a flat, circular sheet, just two cells thick, suspended within the blastocyst, bounded on one side by the original blastocyst cavity (now called the yolk sac) and on the other by the amniotic cavity. Imagine two soap bubbles incompletely separated, joined by a common membrane.

By the end of the second week the pair of bubbles, joined at the germinal disc, is suspended by a thin stalk within a still larger bubble called the chorionic cavity. The whole arrangement buries itself into the uterine wall, stimulating the growth of blood vessels between mother and embryo that will become the placenta, the vehicle by which the growing embryo receives nourishment from its mother, and voids its waste. Everything in this elaborate structure constitutes tissues for the nurture and cosseting of the embryo. All, that is, except for the germinal disc, the minute part of the embryo that will actually become a new human being.

And yet there is nothing about this structure that makes it look like a child. There are no arms, legs, head, skeleton or internal organs. But it is at this stage, during the third week, when the rudiments of a human are sketched out. If the germinal disc is regarded as a map, a flat representation of a three-dimensional sphere, it is possible to see how a human might emerge. For my daughter, as with all other babies, the geography of the human form begins to emerge in the third week, when a furrow appears to plough itself across the top layer of cells in the germinal disc. This is the primitive streak, the Greenwich meridian of the early embryo.

The layer of cells in which the primitive streak forms is the ectoderm – the tissue that will eventually become the skin, nervous system, and much else. The lower layer of cells, or endoderm, continuous with the yolk sac, is the future digestive tract. At the beginning of the third week, cells from all over the ectoderm flow towards the primitive streak like water drawn by a weir, pouring over its edges, cascading downwards to collect between the ectoderm and endoderm.

This distinctive process, called gastrulation, creates a third, middle layer of cells, the mesoderm, between the ectoderm and the endoderm. This new layer of cells will become the muscle, skeleton and internal organs of the new human being. As this process continues during the third week, the primitive streak starts to shorten, leaving a trail in the underlying mesoderm in the form of a tube of tissue. This tube is the notochord, a stiffening rod that will become the backbone.

Towards the end of the third week from fertilization, after the primitive streak has disappeared, another furrow starts to form in the ectoderm, immediately above the notochord. Indeed, this new groove cannot form unless the notochord exists, because chemical signals secreted by the notochord cells are partly responsible for engendering this new structure. This is the neural groove, which will eventually become the spinal cord and the central nervous system.

Meanwhile, knots of mesoderm start to coagulate, in pairs,

one of each pair on each side of the notochord, like rows of poplars lining a French country road. This coagulation starts at the front of the embryo – the region that in the weeks to come will form the base of the skull – and progresses back towards the tail end. These knots are the somites, the segmented 'muscle blocks' seen in all backboned animals, from which derive the vertebrae and other bones, the muscles, and other structures of the body, including the limbs. By the end of the third week, the brain begins to form as two patches of ectoderm on either side of what will become the front end of the neural groove.

The third week is a time of radical transformation, with the creation of a third layer of cells, and the formation of the tissues that will become the backbone, muscles, spinal cord and brain. But these changes are as nothing compared with the frenetic fourth week of gestation, perhaps the busiest week in the life of any human being. This fourth week begins with the process of neurulation, in which the edges of the neural groove grow upwards and inwards, meeting in the middle to form a closed tube. While the neural tube is closing, parts of the paired somites lying on either side grow inwards to surround the neural tube and notochord. While the neural tube is forming, something quite magical happens. A group of cells in the parallel edges of the neural tube migrate from the ectoderm of the neural tube, where they originate, and embark on a mission to transform parts of the rest of the embryo. As these cells, known collectively as the neural crest, travel through the body, they interact with a variety of rather ordinary cells and make them into something special. Much of the skin, the sense organs, the bones of the face, and many other structures owe their origin to the neural crest.

The majority of this activity – the formation of somites, the notochord, the neural tube, and so on – is concentrated in the upper surface of the embryo: in the ectoderm and the mesoderm immediately beneath it. By comparison, the growth of the endoderm is slow. This disparity puts the embryo under such strain that it can no longer remain a flat disc. The burgeoning upper surface curls round, downwards and inwards, surrounding the endoderm, the edges meeting beneath. In this way, the embryo folds itself round the stagnating yolk sac to create a gut tube, and a three-dimensional creature is realized from a two-dimensional map. In engineering terms, the embryo becomes a system of nested tubes rather than layers, making possible the development of those systems of tubes without which life is impossible – the heart, the major blood vessels, the stomach and intestines, and so on.

By the end of the fourth week a human embryo is about the size of a garden pea, and has acquired the rudiments of limbs, kidneys, and eyes, and the very first outlines of a face. It still has much to achieve, but subsequent events are essentially elaboration on a pattern laid down in the first four weeks after conception. Directing this fervid activity is the genome, the agent that creates form from the formless. That the genome sets in train events which create a recognizably human embryo from a single zygote in less than a month is indeed remarkable – yet many questions remain to be asked.

For example, if speed is of the essence, why is the process of human development quite so complicated? Would it not be more efficient to grow a baby from a ball of cells directly, without first flattening a spherical blastocyst into a germinal disc, only to roll it all up again later? Might there not be simpler ways to make an embryo? Possibly, but that isn't the point: the genome has a history, and embryos are not created anew each time, as if from scratch. The genome is inherited, passed down through a chain of ancestors unbroken since the dawn of life, more than 3 billion years ago. So, as well as creating each new human, the activities of the genome reflect the evolutionary history of the human species as a whole. In the dance of its formation, an embryo is paying homage to the deeds of its ancestors. And some of those deeds were done an extremely long time ago.

More than 300 million years ago, our fishy and amphibian ancestors were creatures whose eggs, laid in water, were small and contained little yolk. That is true of the embryos of modern frogs, for example, and the yolk is entirely subsumed within the embryo. Because of this, frog embryos develop more or less directly from a ball of cells: at no stage is the ball rolled out flat to create a germinal disc which must be rolled up again later. Nevertheless, the presence of yolk – which, as in human embryos, is associated with the endoderm – makes the animal's endoderm much slower to develop than the ectoderm.

When our remote ancestors, the earliest reptiles, started laying eggs with hard shells, on land, the embryos had to be supplied with enough yolk to feed them through a long period of incubation. In the egg of any reptile or bird, the volume of yolk is so great that the embryo is tiny by comparison, a flat disc of cells pressed up against the yolk like a lentil stuck to a watermelon. So it was with our egg-laying, reptilian ancestors. One lineage of reptiles evolved into mammals which nurtured their young in a womb. This event happened perhaps 100 million years ago, when our ancestors were small, rat-like creatures scuttling around in the shadows of the dinosaurs.

The yolk sac of a human embryo is a vestigial structure, nothing like the enormous yolks of birds or reptiles. At a very early stage in its development, the human embryo becomes implanted in the lining of the womb, where it produces blood vessels which tap into the maternal circulation, allowing it to feed directly from the mother. Yolk is therefore unnecessary – and neither, therefore, is the requirement that human embryos should be flattened into a disc, to make the most of the space between a large yolk and a hard shell, because neither yolk nor shell have existed in the human lineage for millions of generations. And yet, even now, each human embryo is rolled out to form a germinal disc, reptile-fashion, before rolling up again. There is no greater argument for the gradual evolution of humans than the continued existence of these ancient vestiges.

An echo of an even more distant time in our evolutionary history is found in the formation of the notochord, that rod-like structure which forms beneath the primitive groove in the third week after fertilization. In backboned animals, the notochord is a place-marker for the backbone that eventually replaces it. Long before a human baby is born, its notochord is replaced by vertebrae – the bones of the spine – and the notochord itself survives only in the disks of spongy tissue that separate and cushion the individual vertebrae.

A notochord is also possessed by creatures that never have a backbone at any stage of their life: the shy, rock-pool creatures known as tunicates or sea squirts, as well as superficially fish-like animals called lancelets. Most sea squirts lack a notochord as adults, but this structure is found in the tadpole-like larvae of many sea squirt species. In lancelets, and in the most primitive fishes, the notochord is retained throughout adult life. All animals with a notochord also have a hollow spinal cord, lying above the notochord – without which the spinal cord could not form. The creation of a hollow nerve cord from the rising and coalescence of twin wave-crests of tissue is common to all these creatures. Taken together, the notochord and spinal cord are a distinctive marker, demonstrating a close kinship between backboned animals, sea squirts and lancelets.

Lancelets are vaguely fish-like in outline, and are our closest relatives in the world of invertebrates, almost – but not quite – fishes.[2] In many ways, lancelets seem like fishes only half-made: they have a notochord, a spinal cord and somites, but no bones,

11

fins, eyes, ears, teeth or skin. They have virtually no brain and not much of a head. Lancelets always remind me of the last figure in Shakespeare's seven ages of man − *sans* eyes, *sans* teeth, *sans* taste, *sans* almost everything. But the key difference between lancelets and ourselves is the neural crest, a kind of tissue found only in backboned animals, whose influence on other tissues creates much of the skull and skin, and many parts of the heart, limbs and sense organs. The lancelet is living proof of the importance of the neural crest, for it is as close to a backboned animal as it is possible to be in the absence of a neural crest.

Lancelets are the closest relatives of backboned animals that exist in the modern world. The neural crest evolved in the earliest ancestors of backboned animals, after their lineage had become distinct from that of lancelets. Given that fossils of backboned animals − very primitive fishes − have been found in rocks laid down 530 million years ago,[3] the common ancestor of lancelets and backboned animals must have lived even earlier. The common ancestor of backboned animals, lancelets and sea squirts − that is, the first creature with a notochord − must have lived even earlier than that. In the gestation of any individual human, the notochord forms three weeks after fertilization and disappears as a discrete structure before birth. The neural crest cells appear in the fourth week, and soon disperse to work their special magic throughout the body. As episodes in a single pregnancy, they are fleeting; as markers of our heritage, they are incomprehensibly ancient.

Within the headlong rush of the first few weeks of the development of a single human embryo, it is possible to make out still fainter echoes of the story of human evolution. In the fourth and fifth week after fertilization, the neck and lower facial region of the newly folded embryo start to pucker into a series of folds and ridges − the pharyngeal arches. There are five pairs of these ridges, one member of each pair on each side of the embryo. The fates of the pharyngeal arches are many and varied. Tissues from the first arch – nearest the head end of the embryo – become the jawbone, the muscles concerned with chewing, and the hammer and anvil bones of the middle ear; the second arch gives rise to the stirrup bone in the middle ear, the hyoid bone, parts of the tongue, the muscles of facial expression, and so on. Tissues from successive pharyngeal arches contribute to the thyroid cartilage ('Adam's apple'), the larynx and part of the aorta – a major blood vessel that emerges from the heart.

The pharyngeal arches form anew in each human embryo. At the same time, they are of great antiquity, dating from a time when our remote ancestors were simple sea creatures that fed by straining particles of food carried in sea water. They sucked water in through their mouths and expelled it through a series of clefts in the pharynx, retaining any particles with sieve-like organs which covered the clefts. This mode of feeding can be seen today among sea squirts and lancelets. The young of primitive, jawless fishes called lampreys are also filter feeders, even though adult lampreys have abandoned this peaceable habit for a life of parasitic predation. The habits of young lampreys represent a memory of a time when, between 400 and 500 million years ago, the extinct relatives of lampreys were filter feeders even as adults.

Pharyngeal clefts can also be seen in obscure sea creatures called acorn worms, which are more closely related to star-fishes and sea urchins (known collectively as the echinoderms) than they are to backboned animals, or even to sea squirts and lancelets. Some extinct echinoderms had structures very like pharyngeal clefts, although no modern echinoderm has anything similar; and the last echinoderm with structures resembling pharyngeal clefts died out more than 300 million years ago.

The presence of pharyngeal clefts in such a wide range of animals, from human embryos to the long-extinct cousins of starfishes, implies that this characteristic arose in the common ancestor of all these creatures – long before the evolution of the notochord in the more exclusive common ancestor of sea squirts, lancelets and backboned animals. This creature must have lived at least 550 million years ago.

It is a source of wonder that the attributes that define and justify our everyday humanity – our faces, our expressions, our voices, even the beating of our hearts – all stem from the feeding gear of some emotionless, expressionless animal which dwelt in a rock pool more than half a billion years ago. Such is the depth of our heritage. Even so, it is important to remember that the pharyngeal clefts in human embryos resemble the gills of a larval lamprey only inasmuch as a caricature resembles the real thing. The pharyngeal arches in human embryos are never used in feeding. (Indeed, they are never even perforated.)

All these vestiges do is remind us that individual development has an evolutionary history, too. The genome, which is ultimately responsible for this development, cannot produce a human being in what would seem to us a simple and direct way, without reference to the passage of its own evolutionary adventures over billions of years. The germinal disc thus represents not just a stage in the development of an individual human, but a stage in the evolution of humanity and of life as a whole over more than 3 billion years.

Passed down from generation to generation, the genome is the common thread that runs through all the organisms that have ever existed on our planet. But the passage of the genome from parent to offspring is not so assured that mistakes cannot be made. Sometimes these mistakes lead to stillbirths, or to monsters. Not all mistakes are so destructive, and their accumulation over countless generations leads to variation: variation between individuals, and between different species. Variation is the staff and life of evolution. Without variation, change cannot happen, and it is in the heritable genome that any change is cemented and memorialized. Because the genome has been evolving for such a long time – about a quarter of the age of the Universe – the accumulation of variation has led not only to the otherwise inexplicable richness of the development of individuals, but to an amazing abundance of different species.

The exact number of species that have ever existed is unknowable, but likely to be very large indeed. Current estimates of the number of species living today – an instant in the history of the Earth – average in the low tens of millions.[4] People in this environmentally depleted and increasingly urbanized world might find the scale of this diversity exceedingly hard to grasp. For those of us unable to go hiking in Amazonia, the closest we can get to the true biodiversity experience is to seek out an unreconstructed museum created at the triumphal height of the Victorian age, when London could command every corner of the Earth, and when the reaction of curators to the impending doom of species was not to conserve or breed the remaining individuals, but to kill and embalm them for 'posterity'.

One such museum – and a particular favourite of mine for its undisguised contempt for postmodern, politically correct squeamishness – is the Rothschild Zoological Museum, in the small town of Tring, a short drive north-west of London. It was founded by Lionel Walter Rothschild, later Lord Rothschild (1868–1937), an enigmatic and eccentric scion of the wealthy and influential Rothschild banking family.[5] It is lovingly maintained in its Victorian state by the trustees of the more famous Natural History Museum in South Kensington – who have, thankfully, done their best to maintain the ambience of this spectacular tribute to the art of taxidermy. Rothschild started collecting natural history specimens as a boy. Thanks to a monomania backed by immense wealth, he had, by his death, amassed the greatest natural history collection ever made by one man.

The ranks of mounted trophy heads remind you that this collection was made at the zenith of the British Empire. All animal life is there in its bewildering variety – an Alexandrian library of the possibilities of nature, from elephants to elephant seals, pipefishes to pangolins. There is an impressive collection of stuffed domestic dogs, a great polygonal case of humming-birds, and a lion at the head of a staircase who would look commanding were he not quite so threadbare. There are birds in marvellous profusion: orioles and oropendolas, kittiwakes and cormorants. There are giraffes, gorillas, giant tortoises, a belfry of bats and a room devoted to zebras. There is even a case of fleas, each dressed in costume (and if that in itself is not a testament to eccentricity, I don't know what is). Many of the animals on display in the museum are now extremely rare in the wild, if not actually extinct. There is a case of endangered and extinct birds, in which the Carolina parakeet rubs plumage with the passenger pigeon, and all hold court before that epit-ome of extinct birds, the dodo.

The collection is so rich and so diverse that first-time visitors to Rothschild's museum are overwhelmed. Even seasoned visi-tors come away having seen something new in the seemingly unchanging cabinets. The collections may contain animals that visitors will never have dreamed existed. Rothschild clearly spent a lot of time making his collection comprehensive. For example, you might be unfamiliar with the pangolin – an animal armoured with triangular, overlapping scales, so that it looks like a gigantic pine cone. Well, the museum has a whole case of them, showing pangolin species large and small. You come away from the building with the impression that evolution has produced this cornucopia of diversity with a degree of insouciance that borders on effrontery.

The scale of the diversity of nature, especially in the tropics, had a lasting impact on another Victorian collector, the young Charles Darwin, whose five-year voyage on the *Beagle* instilled in him the germ of what was to become his theory of evolution by natural selection. Before Darwin, the variation of nature was held to be the visible sign of a fallen world. His great insight was to see that variation was not an irritating consequence of imperfection, but the very engine of change and the ultimate source of the diversity that confronted him. Although Darwin had little clear idea about precisely how animals and plants passed on their traits to their descendants, he grasped an essential quality of the genome – its continuity between past and future, with diversity a simple consequence of the genome's antiquity. It is no coincidence that the only illustration in the *Origin of Species* depicts evolution as a tree, with a root and stock giving rise to ever-bifurcating branches and twigs.

In any tree, whether growing in a forest or used as a metaphor for evolution, the branches are not divided into neat segments. In a real tree the trunk leads to boughs, branches and twigs in smooth continuity. Any decision about where the trunk ends and the branches begin is likely to be arbitrary. So it is with the metaphorical tree of life. Modern evolutionary biologists are deeply concerned with the problems posed by attempts to demarcate one species from another: the sheer difficulty has ramifications in fields as rarefied as the philosophy of taxonomy, and as practical as drawing up policies for the conservation of wildlife. The problem of imposing lines on diversity is met in another guise in the creation of form from the formless, the development of a new individual from a zygote. Those committees whose unenviable brief it is to articulate the ethics of research into human reproduction often end up trying to answer an impossible question – at what point does a new life begin?

I could argue that the formation of the germinal disc marks the start of a new life, for it is at that point that one can clearly discern the germ of a new human amid those structures destined only to support its growth, and which will be discarded later. However, I should not like to press the point, and it would be just as great a mistake to cleave too rigorously to the idea that a new life begins precisely at conception. To be sure, conception is the moment when a new and unique genetic constitution is initiated – the combination of genes from both mother and father that will frame the new child's physiology and, to an extent, their character – but this step was itself long in preparation.

Even before fertilization, maternal genes in the egg cell act to dispose various substances around the cell that influence the course of development once fertilization has taken place. Maternal genes direct the geography within which the embryo subsequently finds itself, determining its north, its south, its east and west. Maternal genes are responsible for igniting the new genome and directing its first steps, regulating the activities of genes in the zygote which, in turn, dictate the basic ground plan of the embryo. These zygotic genes would not work at all but for maternal genes which are active even before fertilization. And in this sense we can see that the conception of my daughter, although unique as an instance, did not spring from nothingness but was the continuation of an ongoing process, part of the skein of life.

Once an embryo is formed, almost the first thing it does is prepare for the generations to come. This can be seen as early as the second week after fertilization, when the embryo is still in the germinal-disc stage. At that time a small group of cells in the ectoderm detaches itself from the embryo and migrates to the yolk sac. These are the primordial germ cells – the ancestors of the eggs or sperm of the adult. In the fourth week, when the embryo has folded in on itself, the primordial germ cells return to the embryo, coming to rest in two patches on the back wall of the body cavity. Their journey complete, these prodigal cells stimulate the development of structures that will eventually become the sex organs – the ovaries of the female, the testes of the male. The implication is that the bodies of men and women prepare for the engendering of their own offspring long before they are born themselves, or indeed before they were any more fully formed than germinal discs inside the wombs of their own mothers.

Such themes of continuity with the past and future make it very hard to point to any event in the career of an embryo and declare that it marks the origin of new life. Does life begin at conception, with the germinal disc, or at birth? In this light, it could be argued that the life of any individual begins not with the creation of their unique genome at the point of fertilization, but with the conceptions of its parents, or of any of its progenitors to an arbitrarily remote degree. The debate is fuelled by the misconception that there is a clear dividing line between the lives of individuals, when what actually exists is a strand of continuity which runs back to the beginning, and, by extension, into the future. Our parents give us all the faults they had, but we see in the development of individuals more than the likenesses of our parents: for in such development we can hear the echoes of our evolutionary past.

2

Ex Ovo, Omnia

The question of what it is that produces form from the formless has captivated and fascinated people since antiquity.[1] Even today, the arrival of a newborn baby confronts us with the marvellous, made all the more so by the fact that birth follows a routine which verges on the casual. It is only natural that human beings should have wondered at the unseen and miraculous forces of reproduction since time immemorial: each and every creation myth is an expression of this wonder.

The earliest attempts to answer the question of the origin of form centred on the egg. From apparently formless eggs come a multitude of different creatures, but the origin of these creatures from within an apparently homogeneous medium, the contents of an egg, were quite mysterious. This mystery has attracted a wealth of mythic and cultural associations. Gods and demons in myths the world over are seen to hatch from eggs. Even modern physicists – of all scientists, the most sensitive to the power of myth – talk of the Universe hatching from a 'cosmic egg', when they freely admit that the first instants following the Big Bang are perhaps forever inaccessible to theory or observation.

The first-century Roman scholar Pliny the Elder was entirely seduced by the pervasive mythology of eggs, as he was by every other facet of the natural world, no matter how dangerous the observation (he died from the effects of smoke inhalation while making notes on the same eruption of Mount Vesuvius in 79 CE that buried Pompeii and Herculaneum). Pliny blithely mixed fact and fancy to such a degree that he was described by one later scholar as that 'voluminous, industrious, unphilosophical, gullible, unsystematic old gossip'.[2] In his *Natural History*, Pliny gives an account of a magical egg laid by serpents. Every summer, he wrote, it is possible to witness an egg formed from the saliva and sweat of a writhing bundle of snakes. The egg squirts out of the knot like a champagne cork – 'the serpents when they have thus engendered this egg do cast it up on high into the aire by the force of their hissing, which being observed, there must be one ready to catch and receive it in the fall again'.[3]

The catcher must then leave the scene as quickly as possible, preferably on horseback, for fear of pursuit by the angry snakes, who can only be stopped by a body of water. But the egg, if dropped, 'will swim aloft above the water even against the stream, yea though it were bound and enchased with a plate of gold'.[4] The historian of science Joseph Needham, who quoted this passage in his *History of Embryology* (1934), suggests that whereas the writings of Pliny (in the seventeenth-century translation by one Philemon Holland) are deficient in accuracy, their entertainment value is sufficient compensation.

Aristotle (384–322 BCE) was more clear-headed than Pliny, and can be fairly said to have been the 'father of biology'. He was one of those people who thinks seriously about things everyone else takes for granted. In this case, Aristotle wondered how and why women gave birth to babies with such assurance and regularity, each child so exquisitely and minutely formed, where nothing had before existed. Where did babies *come from*?

21

Any expectant parent may wonder at this same everyday miracle, yet it was Aristotle who, as far as we know, first asked about the forces that made babies possible. He essayed a classification of living things, and considered how they came to be the way they were, and, significantly, how organisms acquired their form as they developed.

Given their cultural importance, eggs were a good place to start an investigation into the origin of form. Methods of incubating hens' eggs had been developed by the Egyptians in antiquity, so the study of eggs posed relatively few practical difficulties: simply open eggs at various stages of incubation and describe the contents. Aristotle, presumably familiar with this ancient but reliable technology, made a distinction between animals which laid eggs and those, such as human beings, which gave birth to live young. He supposed that human babies formed when unshed menstrual blood was sparked into life by the semen, the active male principle. Aristotle's views accorded with ancient Greek tradition that ascribed all inheritance to the male line: women were simply vessels for nurturing the germ of the male.

As with so many of Aristotle's pronouncements, it is a measure of the man's authority that subsequent investigators chose to take his hypothesis as an established truth without exploring it for themselves. However, there is a tale – hopefully apocryphal – of a proposal to put his idea to the test, in which Cleopatra ordered the killing and dissection of female slaves in various stages of pregnancy. Apart from that, Aristotle's ideas had to wait more than 1,500 years before they were tested. When this finally took place, it was found that not only had Aristotle been quite wrong about menstrual blood, but the dichotomy between laying eggs and bearing live young was false all along. In fact, all animals – whether superficially egg-laying or live-bearing – came from an egg.

This challenge to Aristotle came in 1651 with the publication of a book called *Exercitationes de generatione animalium* ('On the Generation of Animals') by the English physician William Harvey (1578–1657). At that time, the question of the origin of form was referred to as the problem of generation. Harvey's work on generation came at the end of an illustrious and sometimes controversial career. His life started auspiciously. He was born into a well-to-do family at the height of the reign of Queen Elizabeth I. He took his degree at Cambridge in 1597, after which he went to Padua in Italy to study under the famous anatomist Hieronymous Fabricius (1537–1619). Fabricius was interested in the question of generation, and based his theories on detailed studies of eggs and chickens. Finding the reproductive tract of hens to be exceedingly convoluted, Fabricius reasoned that any semen from a cockerel would find it difficult to navigate its way towards the eggs. Even if this were possible, the semen would hardly have been able to influence the egg – which, by the time it has completed its formation inside the hen, already has a thick shell. Fabricius suggested, therefore, that the male principle does not contribute to the chick in any material way, but acts as a kind of wake-up call (just like a cockerel), using the energy and vitality of the male to stir the otherwise quiescent egg into development. Harvey was intrigued by his teacher's views, which sparked a lifelong interest in the question of generation.

Harvey returned to England in 1602, and soon married a daughter of one of Elizabeth's physicians. This royal connection, combined with his extraordinary energy and intelligence, led to a busy and profitable career as a physician. One of the perks of his royal association was access to the best seats at the Globe Theatre, in the era of Shakespeare. The world premieres of some of the world's greatest dramas may have been witnessed by one of the Elizabethan world's greatest medical scientists. In

1618, Harvey was appointed physician to the court of James I, a post he retained under the rule of James's successor, Charles I. Ten years later, he rose to academic fame with the publication of his ideas on the circulation of the blood.

Harvey's life, so blessed thus far, took an ill turn with the outbreak of the English Civil War. After Charles I was defeated at the Battle of Edgehill in 1642 (at which Harvey was present), the court, including Harvey, fled to Oxford, but despite the upheaval Harvey settled back into academic life as the Warden of Merton College. While in Oxford, Harvey met the anatomist Nathaniel Highmore (1613–85), who shared his interest in generation. The two men worked together at various times over the next four years. Highmore was an accomplished exponent of the relatively new science of microscopy, and it has been suggested that Harvey's ideas were shaped by the revelations afforded by the microscope.

Respite from war was short-lived. Oxford fell to the forces of Oliver Cromwell in 1646. The King fled once again but was eventually defeated, and he was executed in 1649. Life for former courtiers was doubtless uncomfortable; most of Harvey's private papers were destroyed by Cromwell's troops. Any other septuagenarian ex-courtier living under a regime which viewed him as a potentially subversive element would have gone into discreet retirement, but Harvey seems to have been spurred into yet greater activity. He still saw patients, worked tirelessly for the College of Physicians (where he founded and endowed a library) and carried on researching. *Exercitationes* was his swan-song, published when he was seventy-three. It is notable that in that same year, 1651, Harvey's colleague Highmore published *The History of Generation*, containing the first micrographs published in England. Harvey died in 1657, not quite managing to outlive his nemesis, Cromwell, who died the year after.

Exercitationes was a summary of a lifetime's work on generation. It was based on Harvey's work on recently mated does culled from the deer parks of his erstwhile royal patron. Examining the wombs of the dissected animals, Harvey found no fauns coagulating from menstrual blood (as Aristotle had predicted), no trace of semen, nor, indeed, anything that might bear in any way on the question of generation.

Harvey could have regarded the absence of semen as vindicating the views of his old mentor, Fabricius – that semen does not play a direct role in generation. But there was something else amiss. Fabricius had studied the reproductive tract of hens, which – even without semen – are visibly busy with the creation of eggs, plainly visible to the naked eye. The reproductive tracts of deer made as stark a contrast as might be imagined. In contrast to the insides of hens, the reproductive tracts of does, even recently mated ones, were barren and bare. Not only could no semen be found, but there was no sign of anything else that might betray the origin of new lives.

Inasmuch as Harvey found no trace of embryos being formed out of blood, Aristotle's centuries-old view was plainly wrong – but Harvey could offer nothing that might stand in its place. He confessed himself stumped, and a brave and honest confession it must have been, given his distinguished past. But he was as honest an observer as Aristotle had been, and believed the evidence of his own eyes, even though he could not account for this evidence. His contemporaries (including the game-keepers who tended the deer in the royal parks) swore by Aristotle's ideas and insisted that Harvey must be mistaken.

Still puzzled, Harvey embarked on a further series of investigations of the reproductive tracts of does that had been mated days or weeks before, long after the influence of semen must have worn off (if semen had had any effect at all). He discovered, within the wombs of the deer, formless, water-filled sacs that were not present in unmated deer, but whose exact origin was

25

a mystery – presumably the consequences of phenomena too small to see. It is now thought that the sacs observed by Harvey were the very early embryos of deer, implanted into the uterus wall and each surrounded by a translucent membrane, or amnion, but within which no distinct form could easily be discerned. Significantly, although Harvey was unable to establish a direct, causal link between the fact of mating and the origin of these objects, he made the bold, intuitive leap that these featureless sacs were in fact eggs, in every way equivalent to the externally laid eggs of hens. Aristotle's inferred fundamental difference between egg-laying and live-bearing creatures was, therefore, spurious. Harvey came up with the concept of the egg, or ovum, as an example of a primordium, a more general concept that encompassed the eggs of hens as well as the fluid-filled sacs inside dissected does. Both were destined to grow into adults, nourished by the mother – whether directly, in the uterus, or remotely, by the yolk of a new-laid egg.

It is important to remember that Harvey, looking at the eggs of hens or the early embryos of deer, could have made no distinction between them inasmuch as he would have regarded both as primordial – members of the same category. This is in marked contrast to what we now think of as eggs and embryos, which are quite different. Eggs are single cells whose activities are determined by the genome of one individual, the mother. Embryos are more complex objects, usually made of many cells, whose state is determined by the fusion of two genomes, each parent having made its contribution. Harvey could have known nothing of this: in his time, there was no distinction between single and multicellular organisms, because cells were not yet recognized as the fundamental building blocks of organic life that we understand today.[5] Fertilization – the fusion of egg and sperm – was also not understood. In the 1650s, sperm had yet to be discovered. In any case, Harvey had learned from his mentor, Fabricius, that

26

semen played an indirect role in the process of generation, spurring the egg into life by a process generally termed fecundation, without direct contact between egg and sperm.

Although Harvey could not distinguish between ova and embryos – regarding both as the same kind of object – the fact that he was seeing not the actual ova of deer but early embryos probably in the germinal-disc stage, not long after implantation, does not detract from the importance of his insight that all life came from the egg. The title pages of the first two editions of *Exercitationes* showed the enthroned Zeus holding an egg at the point of hatching, an egg from which a parade of beasts is about to emerge. The egg is inscribed *Ex Ovo, Omnia* – Everything Comes from the Egg.[6]

Harvey's insight, remarkable though it was, fell short of a full account of generation. He had shown that everything came from the egg, but he could not account for *how* form emerged, or where it all came from. Harvey speculated that the egg or primordium is truly formless, and that the embryo develops gradually from homogeneous matter by a process he called epigenesis. However, this says no more than that form arises out of nothing by some unspecified mechanism. As a name, epigenesis is a wild-west storefront with nothing behind it. At best, it is an observation of what happens – that is, form emerging from nothing – not an explanation of why it does so. In coining the term, Harvey essentially sidestepped the issue of generation. Harvey was celebrated as an anatomist and physician of great skill and judgement, but despite his efforts the question of the origin of form remained open.

What most caught the imagination of Harvey's contemporaries was his insistence on the egg as a fundamental unit of life. The next generation of anatomists began to think that the form of the embryo was not created from nothing, but was present in the egg all along, just waiting for that vital seminal spark to

prompt it to grow to visibility. This idea (known as preformation – see below) had been hovering in the wings of thought
since antiquity, and many scholars had touched on it as a possible
explanation of the origin of form. One scientist enthused by
these ideas was the distinguished anatomist and microscopist
Marcello Malpighi (1628–94).

The one facet of Malpighi's character that transcended all
others was his determination. He was born a farmer's son near
Bologna in Italy, and had to shoulder the burden of caring
for his siblings after the family was orphaned when he was
twenty-one. This did not stop him from graduating in medicine
at the University of Bologna four years later. Although his
academic career took him all over Italy, Malpighi returned to
Bologna in 1666, where he was to stay for most of the rest of
his life, concentrating on his anatomical studies.

It was here that Malpighi embarked on a one-man voyage
into the microcosmos, making many spectacular discoveries in
the then very new science of microscopical anatomy. He
observed blood cells, establishing the fine connections between
arteries and veins and thus completing Harvey's pioneering
work on the circulation of the blood; he established that the
lungs had a fine structure of tubes and cavities, and were not
simply formless spongy masses; and he established the anatomy
of the spleen and the kidney. Generations of biology students,
even those studying today, can remember having to draw the
network of vessels in the kidney now known as the Malpighian
tubules. But Malpighi was wrenched away from both the microscope and Bologna in 1691, when, at the age of sixty-three, he
was summoned to Rome to be physician to Pope Innocent XII.
Malpighi died in his post three years later (the Pope, however, lasted until 1700, perhaps as a testament to Malpighian
ministrations).

Malpighi's long love affair with his microscope did not make him a hermit. He was a keen correspondent with other scientists in Europe, including those at the Royal Society in London. Now the leading scientific body in Britain, the Royal Society was then a fledgeling institution, for all that its early fellows numbered such luminaries as Robert Hooke and the chemist Robert Boyle.

In common with many microscopists of his day, Malpighi had made a detailed study of the development of the chick in a quest for the origin of form. An adept microscopist, he was perhaps more capable than most of teasing out the very smallest structures, on the limits of resolution, that could yield clues about how form originated. Did embryos really coalesce from nothing, as Harvey had suggested, or did they emerge from preformed germs that might yet be discerned by the keen-eyed investigator? Malpighi published his findings in 1673, as *Disserta-tio epistolica de formatione pulli in ovo* – 'The Formation of the Chicken in the Egg'. Although he described the development of the chick embryo in unprecedented detail, the very earliest stages of generation eluded even him. Given the questions about the role of semen raised by Harvey and Fabricius before him, Malpighi wondered whether the form of the chick might be discerned in unfertilized eggs. He could make out no signs of life in unfertilized eggs, but thought there might be germs of an embryo in eggs that had been fertilized but left unincubated. Working on the very edge of the unseen, Malpighi – good scientist that he was – admitted that he could not rule out the possibility that the germ of the chick might not reside in the egg before fertilization. In other words, he considered seriously the idea that the form of the chick might not have been created anew in each egg by a process of epigenesis as envisaged by Harvey, but might, in each case, represent the instantiation of some eternal pattern residing in eggs, continuous between the generations.

Similar thoughts had occurred to a researcher working in the Netherlands, Jan Swammerdam (1637–80), who had turned his microscope on many things and was, like Malpighi, interested in the silkworm, an animal of great commercial importance to the textile industry. Swammerdam's party-piece was to dissect a silkworm chrysalis, revealing the rudiments of the adult moth to the astonished guests. Today this might not astonish us, but in those days the chrysalis of an insect was regarded as no different in concept from the egg of a chicken. In the context of the time, any knowledgeable person would naturally conclude that the rudiments of the adult moth were in the chrysalis all the time, simply waiting for their opportunity to expand to the level of visibility and, finally, emerge. Swammerdam speculated that if they were examined closely enough, eggs (in the sense of 'primordia') would be found to contain the minutely detailed, preformed parts of the animals destined to hatch from them. If semen had any role, it was merely as an indirect influence, as Fabricius had supposed. Swammerdam suggested that semen might prompt the emergence of preformed parts in the egg by emitting some impalpable essence or spirit, which he named the *aura seminalis*.

The work of Swammerdam, Malpighi and others energized a French theologian-turned-philosopher, Nicolas Malebranche (1638–1715). In 1674 he published *De la recherche de la vérité*, a treatise on the nature of knowledge, generally known in English as *On the Search for Truth*. This work was as vast and comprehensive as one would expect from a disciple of Réné Descartes, and only a small part of it concerned generation. Its influence, however, was enormous because it collected the scattered observations of microscopists and drew from them a general theory that came to be called preformation. According to preformation, the rudiments of every creature had been created by God at the beginning of time, and were simply waiting for their predestined cue to emerge. All the generations of Man would therefore have been found in the ovaries of Eve. Preformation was to remain the predominant theory of generation for the next century.

At first glance this idea seems quite fantastic, but in the seventeenth century it made much sense. The advocates of preformation – and there were many – were accomplished and in many cases brilliant scientists. 'It is impossible to believe', wrote Elizabeth Gasking in *Investigations into Generation* (1967), 'that men of this calibre should have all been deluded into accepting a theory such as this, unless they felt there were some very good reasons for doing so.'[7] One obvious objection to preformation is *smallness*. To suppose that all the generations of Man were wrapped up inside the egg of our ultimate foremother is to propose the existence of structures of near-infinitesimal dimensions, a concept that strains credulity. And yet, the theology, the philosophy and – most importantly – the experimental science of the seventeenth century saw in preformation explanations of many disparate phenomena in the compass of a single scheme. This, then as now, is the mark of a sound scientific theory.

First, the number of generations between ourselves and Eve was, in the seventeenth century, not regarded as infinite or undefined. If the world had been created a matter of a few thousand years ago, as attested in the Bible, and women gave birth at an average age of twenty years, then fewer than three hundred generations could have preceded the publication of Malebranche's *Search for Truth*.

A theological argument – and the one that most motivated Malebranche – concerns the existence of the marvellously intricate worlds of life that had been suddenly brought into view by the microscope, and whose existence had hitherto been entirely unexpected. To Malebranche, and many others, the

31

impact of this discovery was evidence enough for the power and goodness of God, who had conceived the entire world in such detail so that we might discover it.

Perhaps the most important consideration was philosophical. After recovering from the revelatory shock of the existence of the microcosmos, scholars suggested that there was no good reason to posit any lower limit on size. The theory of atoms, let alone that of cells, still lay in the future, and when the first microscopists turned their lenses on nature, they opened up a whole new world of detail in which substances, apparently featureless to the naked eye, were in fact full of structures and complete living forms of great delicacy. What then, the preformationists asked, would prevent an amoeba-sized scientist, equipped with a suitable microscope, from discovering yet more wonders on an even smaller scale? Today, physicists speculate on the nature of matter at such an inhumanly small scale that the fabric of reality itself becomes granular, and wonder if even smaller structures might exist. The justification for these ideas is entirely theoretical, and there is as yet no hope of verifying them by experiment. Much the same could be said of preformation in the seventeenth century: to object to the existence of something because it is far, far smaller than anything in your experience is hardly a scientific criticism.

Malpighi looked at blood and found it to be made of particles we would nowadays recognize as cells. He studied the seemingly formless tissues of lung and spleen and found in them a wealth of detail. Swammerdam revealed the complexity of the silk-worm just before it emerged from the chrysalis. And then a microscopist in the Netherlands turned his microscope on semen, that fluid whose role in generation seemed so ambiguous. The results would come as something of a shock – semen was full of tiny, writhing worms.

3

Unfolding

The identity of the first person to see sperm through a microscope has been a matter of dispute. A strong claim lies with that pioneer of the microscope, the Dutch scientist Antony van Leeuwenhoek (1632–1723), although it was fiercely contested at the time by another microscopist, Nicolas Hartsoeker (1656–1725). Either way, it seems certain that sperm were first seen through a microscope in the early or mid-1670s. Leeuwenhoek did not help his case by announcing his many findings in letters to his friends, who would broadcast the news without necessarily acknowledging their source. This seems to have been the case with sperm. It appears that Leeuwenhoek described sperm in a letter to a friend, the distinguished poet and polymath Constantijn Huygens (1596–1687), who passed the letter on to his equally distinguished but more technologically minded son, Christiaan (1629–95), who would have had more of an idea of what to make of new discoveries made with the aid of the microscope.

Seventeenth-century Holland was synonymous with fine optics in the same way that modern Switzerland is associated with expensive watches. The first telescopes were probably made in Holland, and Galileo used a Dutch telescope – or at least one of Dutch design – to discern the moons of Jupiter in 1610. Less well known is that Galileo may have had a simple microscope at about the same period. The word 'microscope' was coined in 1625, and microscopy became the fashion that no well-connected Dutchman could afford not to follow. Christiaan Huygens was as well acquainted with microscopes as were Malpighi or Highmore, although his fame lay in the discoveries he made with his telescope. By the time he received Leeuwenhoek's letter from his father, Christiaan could claim several important astronomical discoveries, including that of Titan, the largest moon of the planet Saturn.

Christiaan announced the discovery of sperm in 1678, without any attribution to Leeuwenhoek, in a discussion of animals said to arise spontaneously, from putrefaction. In his report, Christiaan describes animals found in semen that are

> formed of a transparent substance, their movements are very brisk, and their shape is similar to that of frogs before their limbs are formed. This discovery, which was made in Holland for the first time, seems very important, and should give employment to those interested in the generation of animals.[1]

The closer you read this important-sounding announcement, the less it seems to say. On the one hand, it suggests that sperm are complete yet very small animals, no different from the other small animals that microscopists were discovering, in pond water, for example. At the same time, the announcement suggests that sperm might be of interest to those studying generation – or, then again, they might not. The role of sperm in generation was still not as clear cut as the role of the egg.

Many of the early microscopists were physicians or anatomists. For a doctor, at a time when infestations and infectious diseases were common and the importance of sanitation was unsuspected, it would have been quite natural to look at sperm and think they were no more than another case of worms. That flies, worms, and so on were spontaneously generated from diseased or putrefying matter was a seemingly obvious deduction in a world in which the stench of decomposition was familiar to everyone, and was assumed to be true by such authorities as Christiaan Huygens. No wonder then, that when sperm were first discovered, they were almost automatically *assumed* to have been interlopers – parasites – with, perhaps, no direct relevance to generation.

This is why the first impulse of many of sperm's earliest observers was to classify them in their own right, as something quite other than the animals they appeared to inhabit. In 1700, the French physician Nicolas Andry de Bois-Regard published a large work entitled *De la génération des vers dans le corps de l'homme* ('An Account of the Breeding of Worms in Human Bodies'), which became a standard work on medical parasitology. One kind of worm had a special place in Andry's affection – the so-called spermatic worm. Andry's views exemplify a general ambiguity about the role of sperm which persisted for more than a century: he was quite prepared to believe that sperm played a dual role, as free-living parasites and as carriers of the preformed embryos of organisms. In *The Ovary of Eve: Egg and Sperm and Preformation* (1997), her sparkling reassessment of preformation, biologist and historian Clara Pinto-Correia suggests that the close association drawn by Andry between infestation and generation would have damaged the prevailing concept of sperm as agents for the spark of human life, and this association would have acted as a strong disincentive to scientists who might have sought in sperm the answers to the great questions of generation.

However, it was clear to Andry and to other observers that sperm were found only in male animals of reproductive capacity and in good health, and that each species had its own variety of sperm. Given this coincidence, some did begin to wonder whether sperm might not be parasites after all, but particles directly concerned with generation. But it was not a simple matter to rule out other possibilities. As early as 1679, Robert Hooke reported the presence of sperm in the testes of a horse, and had failed to find them in immature males; yet he could not discount the idea that sperm were parasites specifically found in the testes of mature males. Because of the ambiguity about the role and nature of sperm – as parasites, as agents of generation, or both – the idea that sperm were organisms in their own right proved an enduring one. As late as 1835, zoologist Richard Owen (1804–92) classified spermatozoa as a distinct order of animal life; the term 'spermatozoa' was coined as late as 1827, by the zoologist Karl Ernst von Baer (1792–1876).

The problems of defining the place of sperm in nature dogged the wider acceptance of what came to be called spermism – the sperm-based idea of preformation – that, whatever else they did, sperm acted as the vehicles for the transmission of inheritance, and that all human generations would have been present in the testes of Adam, and not in the ovaries of Eve. These problems were deepened by several issues that had more to do with the image of spermism than its substance.

The idea that sperm, not eggs, might contain the germs of all future generations was taken up by Nicolas Hartsoeker. In a throwaway remark in a 1694 book mostly about optics, he suggested that were we to have microscopes powerful enough, we might find embryos rolled up inside the heads of sperm. Pictures made subsequently of little foetuses rolled up inside sperm heads were only cartoons lampooning this idea – and yet gave rise to the seemingly unshakeable conclusion that

Hartsoeker and others had actually made and reported such observations, a legend that has since been perpetuated down the ages – by its own memetic regeneration – as an example of how daft and deluded our ancestors must have been to subscribe to preformationism of any kind, whether based on sperm or eggs. To make matters worse, Hartsoeker's apocryphal rolled-up foetus was referred to as a homunculus. This word may seem innocuous enough (after all, it only means 'little man' in Latin), but it had already been appropriated in the literature of medieval alchemy for a person created artificially, by occult recipes or magic. For example, the famous Swiss alchemist Theophrastus von Hohenheim (1493–1541), usually known as Paracelsus, reported a recipe for making homunculi that required a mixture of human semen, human blood and horse dung to be left to putrefy for more than a month, after which the blind stirrings of the homunculi might be observed. No true science of the Enlightenment could retain any shred of credibility if forced to bear the embarrassment of such medieval stenches.

The final knell for spermism was based more on ethics than on science, in particular the propriety of working with human semen as a biological material. Given that masturbation was proscribed by scripture, early microscopists were sometimes less than clear about whose semen they had used for their observations. This reticence became near-total silence after 1715, when an anonymous pamphlet entitled *Onania*, detailing the evils of masturbation and its fearful consequences, achieved wide currency in Europe. After that, no serious discussion of spermism was possible. In 1722, even such a leading exponent of spermism as Hartsoeker publicly renounced this view.

The fading of sperm-based preformationism after 1715 left the field clear for a resurgence of the older idea that preformed embryos were to be found in eggs. This ovism – which ran counter to spermism and in parallel with it – was to become the predominant theory of generation for the rest of the eighteenth century, thanks to three colossal yet complementary talents: a dour Protestant physician named Albrecht von Haller (1708–77); the precocious French entomologist Charles Bonnet (1720–93), discoverer of the phenomenon of parthenogenesis, who went on to become the leading theorist of preformationism; and Lazzaro Spallanzani (1729–99), an urbane Italian priest, arguably one of the finest experimental scientists who ever lived. They were a band disparate in background, temperament and talent, yet their work turned preformationism into a mature discipline, grounded in well-honed theory, supported by rigorous experiment, and unassailable except by developments in experimental science as opposed to changes in theological dogma or ethical outlook.

Haller, the eldest of the three, was widely admired as a physician, although he wasted much energy in fruitless worry – about money, social status, and trying to reconcile his findings with his strict Swiss Protestant outlook. Haller had been a student of the Dutch physician Hermann Boerhaave (1668–1738), who favoured spermism but was sufficiently broadminded to consider all views. This generous spirit made Boerhaave both popular and respected, and his mildly spermist views were initially adopted by his adoring student, Haller, who might have persisted in this view but for the biological bombshell of the 1740s – the discovery of the phenomenon of regeneration.

In the early 1740s, a relation of Bonnet's named Abraham Trembley (1710–84) published his experiments on the freshwater polyp, a microscopic creature with a small stem and tentacles, yet curiously mobile. Opinion was divided on whether the polyp was an animal or a plant. As part of a programme to investigate this, Trembley sliced the creatures into pieces and found that entire, new polyps formed from the fragments. The power of regeneration in such a modest corner of creation earned this organism a powerful name: *Hydra*, after the horrific many-headed monster of Greek myth which startled Hercules by replacing each severed head with two new ones. Trembley's work caused a sensation, and before long other zoologists were busily testing the regenerative properties of various creatures. Regeneration poses an obvious difficulty for preformation. If many complete creatures can be regenerated from a single individual that has been fragmented, the role of the preformed germ is called into question. Questions were indeed raised in Haller's mind, and he started thinking along epigenetic lines. Even though there was no mechanism to explain the origin of form from formless matter, the bare facts of regeneration showed that preformationism was in trouble.

Haller's dalliance with epigenesis did not last. In the 1750s his own studies on chickens and eggs, in the style of Malpighi or Fabricius, led him firmly and finally to preformationism. Close study of the membranes within the egg as it developed convinced him that the yolk was continuous with the skin and gut of the foetus, and that since the yolk could be found in unfertilized eggs, then so too must the embryo.

During Haller's reconversion to preformationism he received a fan letter from Charles Bonnet, who was making a name for himself as a talented scientist with sympathies for pre-formationism. Haller – flattered, naturally – wrote back immediately, starting a lifelong correspondence and friendship. Bonnet's career had begun in earnest just a few years earlier, after he had written to the distinguished entomologist René-Antoine Ferchault de Réaumur (1683–1757), a polymath in the occasional employ of King Louis XIV. Bonnet had read Réaumur's book *Mémoires sur les insectes*, tried a few of the suggested experiments, found them wanting, and sought to communicate with the author. Réaumur, like Haller, was warmed and charmed by the enthusiasm of this bright young man, and under

Réaumur's influence Bonnet made a careful study of aphids. In so doing he discovered what looked to be sure and certain evidence for preformation. The discovery catapulted Bonnet, just twenty years old, to scientific stardom.

As long ago as 1677, Leeuwenhoek had reported to the Royal Society in London that female plant lice, or aphids, could reproduce without males. We now know that aphids, as well as many other animals, can dispense with sexual reproduction and reproduce clonally – that is, they can make genetic copies of themselves. The offspring of this process are always female, and the process itself is called parthenogenesis. The advantage of parthenogenesis is that it allows a species to reproduce with great speed to take advantage of an ephemeral or temporary resource; aphids are very good at this (all the better to consume your roses). What Bonnet found – and his findings remain valid today – was that aphids are so rushed, it seems, that entire generations are telescoped together: the body of a single female may contain the bodies not only of the next generation, but the germs of her grandchildren, even before her own children have been born. Female aphids are living Russian dolls, and offered graphic proof of preformation.

Aphids are tiny, and the strain of trying to discern generations of progeny concealed within their minute bodies undoubtedly contributed to Bonnet's early blindness. This, and his increasing deafness, confined Bonnet to a life of the mind – an enforced concentration which led to his writing a book, *Considérations sur les corps organisés*. Published in 1762, it marked the high-water mark of preformationism.

Bonnet based preformationism on, essentially, two hypotheses. The first stemmed from his observations of aphids, from which he generalized that the bodies of organic beings had stored within them the germs of all beings to come. These germs were not exact miniatures of adult creatures, but a formulation of what Bonnet called their 'essential parts', which would enlarge and become more organized through the process of development. The second hypothesis was rather more fanciful. Bonnet called it dissemination. He supposed that it might not be possible to encase the multitude of future generations in a single organism, and instead that the invisible germs of species might be disseminated throughout air and water, waiting to meet fully formed bodies of their cognate species in which they might lodge and grow.

Bonnet's justification of preformation rested on the seeming impossibility of understanding the initial formation of organisms epigenetically, that is, as if from nothing. Therefore, he reasoned, organisms must have existed for all time (or at least since the divine Creation), either as organized bodies or as germs. In hindsight, Bonnet's notion of essential parts is an exact operational description of the genome, for, in bald terms, the genome can be thought of as the representation of the essential parts of an organism, whose full expression becomes manifest only through development. Crucially, the genome itself is passed down from generation to generation, and in some sense all future generations might be predicated on the genome of our ultimate ancestor, even if the precise form of each generation is conditioned by chance and circumstances rather than divine decree at the moment of Creation. Bonnet's idea of essential parts expresses a continuity of inheritance we now appreciate in the genome. His complementary idea of dissemination, although less justifiable in modern terms, resonates at least emotionally with the concerns of modern environmentalists about how genes might float about in the air, waiting for an appropriate organism to settle in.

Bonnet also had a way of accommodating a recurrent criticism of preformation, that of 'sports', or creatures born with obvious defects such as withered limbs or cleft palates. Surely, detractors asked, no beneficent Creator would have wished disability on a hapless and innocent infant at the very moment of Creation? Bonnet countered that if all generations had been encapsulated in the same moment by the Creator, then there is no reason why all individual variations, including monstrosities, could not have been so accommodated.

Among Bonnet's correspondents was the priest Lazzaro Spallanzani. Spallanzani's father had wanted him to study law. For a while Spallanzani complied, but he was forever distracted by the more exciting prospects offered by the natural world. Eventually he became a priest, and used his ecclesiastical income to fund his passion for nature. Whereas most naturalist clerics were content to observe and make notes, Spallanzani went further: he manipulated nature, created hypotheses and designed experiments to test them.

In his most famous experiments, Spallanzani sought to discover whether semen needed to come into direct contact with eggs in order to fecundate them – that is, to stir them into life. Following the work of Fabricius and others, it was thought that semen had no direct role in fecundation but acted as a source of energy or 'spirit' – Swammerdam's *aura seminalis* – to jolt the egg into development.[2] As an experimentalist, Spallanzani saw that the existence of this elusive fluid might be investigated. At Bonnet's suggestion, Spallanzani chose to do his experiments with frogs, for the simple reason that in the wild male frogs shed semen on the eggs immediately after they are laid by the female. This means that both eggs and semen are accessible just before fertilization takes place, and the effects of the proximity of semen on the properties of eggs can be examined directly. Spallanzani showed unequivocally that semen and eggs had to come into direct contact for the process to work. He proved this in the breach, and in the breeches, by showing that fertilization was impeded if the male frogs were engaged in courtship while wearing little taffeta trousers of Spallanzani's own design.

In another experiment, Spallanzani showed that eggs could be fertilized by frog semen diluted in water to various degrees. But when a mixture of semen and water was progressively filtered to remove any particulate matter, the filtered fluid became less and less potent. If, however, the sediment, deposited on the filter, were immediately redissolved in water, the mixture was as potent as ever.

To modern eyes, the conclusion is obvious: the sediment removed by the filter contains agents necessary for fertilization, and it is at least possible that these agents are sperm, given that sperm are found in the filter and not in the strained and impotent fluid. However, Spallanzani was convinced that the germ of the tadpole was present in the unfertilized egg of the frog; that semen was merely a substance which activated the egg; and that any sperm to be found in the semen were irrelevant and parasitic contaminants. Spallanzani supposed that sperm, as parasitic worms, could infect a new host at a very early stage of development, and might even infect unfertilized eggs. 'It is rather alarming', comments Elizabeth Gasking, 'to think that had Spallanzani really seen the penetration of the eggs by the spermatozoa he would have regarded it as a confirmation of this hypothesis.'[3]

Spallanzani outlived both Haller and Bonnet, and his death in 1799 was soon followed by that of preformation itself. In the first few years of the nineteenth century, the Quaker meteorologist and chemist John Dalton (1766–1844) developed his atomic theory of matter. Derived from a wealth of data on the behaviour of gases and chemicals in general, Dalton's theory showed that matter was not infinitely divisible, and that there was a lower limit on size. Although Dalton was not a biologist, the implication of his ideas on preformation was clear: microcosmos could not forever give way to microcosmos, allowing for multitudes of generations to be telescoped together down to some arbitrary degree of smallness.

Of perhaps more direct relevance to biology – and the fall of preformationism – was the development of cell theory, which came into flower in the early nineteenth century after a long dormancy. Nowadays we are accustomed to the idea that all living things large enough to see with the naked eye are made of a large number of tiny bodies called cells. Human beings consist of trillions of them, most highly specialized to perform certain functions such as nervous transmission or muscular contraction.

Leeuwenhoek saw cells perhaps as long ago as the 1770s. Indeed, sperm are individual cells in the sense that we now understand the term. In the simplest microscopes, cells are seen as featureless blobs, and this is undoubtedly how Leeuwenhoek would have observed them. Rapid advances in microscopy revealed that many cells each contain a smaller, concentrated body now known as the nucleus. The nuclei of plant cells may have been observed in the 1780s, although it was 1833 before the presence of nuclei was documented as a consistent feature of cells. More modern techniques have revealed great complexity within the nucleus, in the other matter, or cytoplasm, that surrounds it, and in the membrane that forms the boundary of the cell.

Most living things live their lives entirely within the frame of a single cell. Pond organisms such as paramecia and amoebae are unicellular, and creatures such as these would have been among the first organisms seen by the earliest microscopists. Leeuwenhoek and his colleagues, on first observing cells, would have had no reason to think that living matter is normally subdivided into legions of tiny compartments – and why should they? To them, paramecia and amoebae were tiny animals, very much alive despite their smallness. According to Aristotle (and to common sense), the status of bodies as living things implied the presence of fully functional systems of organs: a heart, a liver, and so on. When they were discovered, spermatozoa were also supposed to have complete internal organs – even if the head of each sperm did not contain a tiny homunculus. It could not possibly have occurred to a seventeenth-century anatomist, armed with a simple microscope, that the organs of higher animals are composed of cells, each one a discrete living entity, and that it was possible for animals to lead busy, active lives even in the unicellular state, without what we would think of as distinct tissues and organs.

The idea of cells as units of life that were fundamental rather than incidental started with studies of the microscopic anatomy of plants. Hooke, on first noticing regular spaces in slices of cork, named them 'cells' because they reminded him of the discrete, austere rooms of monks. The cells of plants are in general very much larger than those of animals, and have more definite boundaries: stiff curtain-walls of cellulose, rather than the soap-bubble membranes that serve in the cells of animals. This makes plant cells much easier than those of animals to see in a simple microscope. Cells were initially regarded as curiosities of plants, of no relevance to the animal world, and there was then no reason to suspect that cells were any more than structural components, or had anything to do with the creation of life.

With time, more powerful microscopes, and chemical techniques for staining cells and nuclei to make them easier to see, it became evident that all living things – animals as well as plants – were very largely composed of cells, and that these cells were not simply structural elements, like bricks. In 1839 two German biologists, Theodore Schwann (1810–82) and Matthias Schleiden (1804–81), declared that cells were the fundamental particles of organisms, in the same way that atoms were the elementary particles of matter. All creatures were made out of cells. Either they consisted of a single cell, as with paramecia and amoebae, or they were made of many cells stuck together, as in humans.

Schwann and Schleiden's idea came to be known as the cell theory. It did not change the world overnight, because it threw up a number of other issues in its wake. What were cells made of, and what was their origin? The answer to the first question led to the development of the idea of protoplasm (the jelly-like substance from which cells are made) as the fundamental state of living matter, and yet it was unclear how protoplasm originated from the inanimate world – if, indeed, it did. The answer to the question of where cells came from proved both frustrating and inedifying: cells came from other cells. Organisms do not grow by accumulating cells from some external source, but by making more cells from within. Cells were found to be capable of dividing into two, and dividing again, each daughter cell growing until it was the size of the original, and then dividing further – or specializing to become a component of brain, or liver, or muscle, or bone. This division – or fission – was what passed for reproduction in single-celled creatures such as amoebae.

The theory of cells as fundamental units of life put paid to the stupefying infinities of preformation by establishing a lower bound on organic smallness, but it did not in itself solve the problem of the origin of form. Cell theory still left unbridged a great divide between the simplest cell and the most complex non-living matter, to the extent that protoplasm was thought to be a distinct and special substance, containing a vital spark unseen in the world of the inanimate. Nevertheless, cell theory advanced biology by finally opening the way to a theory of generation which did not sidestep the issue by booting it back to the Creation, as Bonnet and Spallanzani had done. If plants and animals are made of collections of cells, each kind more or less specialized for a certain function, could there not be cells that were specialized for generation? Once this realization dawned, spermatozoa were seen in an entirely new light, as cells of the host body, specialized for the task of generation.

As the single-celled representative of a multicellular male, it made sense that the sperm should have a unicellular, female counterpart. In one sense, Harvey, Swammerdam and Malpighi had been right all along in their insistence on ova as the primary seat of generation. But the absence of cell theory, and of micro-scopes consistently powerful enough to resolve animal cells with clarity, left them without the tools necessary to draw a distinc-tion between unicellular eggs on one side, and multicellular embryos on the other, and without which both were regarded as indefinite 'primordia'.

The decisive result came in 1828, when von Baer, a year after he coined the term 'spermatozoa', described the human ovum as a single cell, an austere room in which no space could be found for preformed germs – neither physically nor concep-tually. It became clear that all ova are single, indivisible cells, whether they are very small, like the ova of human beings or of the deer studied by Harvey; or very large, like the eggs of hens studied by practically everyone since antiquity. Once that was realized, the search for nested generations of preformed embryos was finally exposed as futile. Harvey was right – form emerges from nothing, and everything comes from the egg.

4

Revolution

If Albrecht von Haller enjoyed receiving letters from Charles Bonnet, he dreaded any correspondence from a young and very combative Berlin-born physician named Caspar Friedrich Wolff (1734–94). Exchanges between Haller and Wolff descended into a kind of epistolary dogfight, the effect of which was particularly draining for the older man, who could only be soothed by emollient expressions of concern from Bonnet. In 1759, just three years before Bonnet published his *Considérations sur les corps organisés*, Wolff produced a thesis with the ambitious title *Theoria generationis* ('Theory of Generation') in which preformation was comprehensively demolished in favour of epigenesis. The fact that Wolff had dedicated his work to Haller only heaped on the insult, but given the dominance of preformationism, Wolff's ideas initially had very little impact. The only notices Wolff received were sharp criticisms from both Haller and Bonnet. But before he could press his point still further, Wolff was drawn into a dispute of a more lethal kind. In 1756, Prussia and Russia became embroiled in the Seven Years' War (1756–63), and Wolff, newly qualified as a doctor, was posted to a military field hospital. At the end of the war

Wolff was unemployed, and supported himself by giving private lectures while he looked for a job. His radical ideas met only with silent disregard, and it was not until four years later, after intense lobbying by his friend, the mathematician Leonhard Euler (1707–83), that Wolff found an academic position – at the St Petersburg Academy of Sciences.

In St Petersburg, Wolff settled down to a life of intense scientific productivity. From his embryological studies he evolved a theory of 'germ layers' decades before Ernst von Baer, who is usually credited with the discovery. The germ layers are the primary layers of cells – ectoderm, mesoderm and endoderm – from which an embryo is made. Wolff also developed an idea of plant anatomy in which the specialized organs, such as petals and sepals, could be viewed as modified leaves: an idea given more rounded formulation in 1790 by Goethe. At the time of his sudden death from a brain haemorrhage, Wolff had been working on a 'theory of monsters' in which he had planned to systematize his ideas about shape, form and development in animals, anticipating by a century William Bateson's book *Materials for the Study of Variation* (see Chapter 6). Yet Wolff's own work was largely ignored in his own lifetime. The reason was *Theoria generationis*, the foundation of Wolff's reputation (or lack of it), which seemed to have dogged him throughout his career. The basis of the work was the embryology and development of the chick. By that time, the embryology of the chick was well known, having been studied by Haller, Malpighi, Fabricius, and many others, right back to Aristotle, and the various investigators disagreed only on minor points, at least in terms of physical description or observation of the embryo once it had reached a stage of development when it could be clearly seen.

The battles lay in interpreting the earliest stages of development, on or beyond the edge of visibility. Preformationists argued, of course, that all the various parts of the embryo were there from the beginning, even in the unfertilized egg, and simply expanded until they were discernible under a microscope, then a hand lens, and finally with the naked eye. Wolff weighed in with the following disarming observation: 'nobody has yet discovered with the help of a stronger lens parts not also visible with the help of a weaker magnification'.[1] His claim was a slap in the face for the preformationists: the implication of his statement was that it is not necessary to track preformed parts down to infinitesimal size using ever more powerful lenses, as the parts will not be found. The various parts of an embryo are formed and organized on a stage that is visible to the experimenter. They do not inflate fully formed from an ineffably small state. Wolff told it as he saw it and, as such, found himself playing the part of the small boy who demonstrates the exiguity of the Emperor's fine new wardrobe.

To say that Wolff exposed preformationism as empty would be an oversimplification, because it worked very well as a scientific theory in the sense that it explained, with elegance and force, a great deal of the evidence then available. One of the oldest arguments for preformationism was that of the interdependence of parts. It seems obvious that animals are highly integrated beings, in which all the various organs work together so harmoniously that it is impossible to imagine that any function might be possible were any one of these organs missing. Were we deprived suddenly of heart, liver or lungs, we would die instantly. Following this same logic, the preformationists argued that this was as true for an animal during the course of its development as in the adult state. To have organs developing at various times, willy-nilly, out of formless matter, as Wolff described in *Theoria generationis*, was simply nonsense to a preformationist. However, the idea of the interdependence of parts breaks down when the evidence is examined uncritically: when you look at the very earliest stages of embryonic development in a chick, the heart makes a dramatic early appearance, blood-filled and beating, before any other organs can be seen. The preformationists strained both their eyes and all their credulity in maintaining that all the other organs must have been there, but were too small to be seen, or transparent, or of an indeterminate, liquid consistency. Wolff's own work convinced him that the vessels and the other organs of the chick could not be present, even in some occult form, but were later formed from homogeneous matter. And yet the heart still beat, alone and into the void.

Apart from observing that the embryo appeared to form from nothing, Wolff, like Harvey before him, was unable to explain how the various parts coagulated from the indeterminate jelly of the early embryo: 'We may conclude that the organs of the body have not always existed, but have been formed successively – no matter how this formation has been brought about,' Wolff wrote. 'I do not say that it has been brought about by a fortuitous combination of particles, a kind of fermentation, through mechanical causes, through the activity of the soul, but only that it has been brought about.'[2] In other words, he had no idea about the nature of the agency that brought form out of the formless. When pressed, Wolff would give this unknown entity a name: the *vis essentialis*, the life force, whose nature remained vague. Was it a physical entity, or a force that drove development, akin to the gravity that binds matter and motion, or was it perhaps some force innate to living matter? Bonnet ripped this idea to shreds: gravity is one thing, but even if the *vis essentialis* were some generalized force, it would still not explain why creatures develop along their own peculiar trajectories. If pangolins and petunias are moulded by the same *vis essentialis* out of the same formless matter, why do they continue to look so different? Preformation was still a better explanation, for all that it devolved all unanswered questions to the Creator. On the other hand, the *vis essentialis* might not be an entity at all, in the sense of a separate force that acted on living matter. For Wolff, it could simply be a consequence of the essential nature of life, without cause or need of explanation.

Wolff's ideas of the essential nature of things may seem highly unsatisfactory to us now, as they evidently did to the pre-formationists. But they struck a chord with a philosophical tendency that was especially strong in Germany from the mid-eighteenth to the early nineteenth century. This was the school of *Naturphilosophie*, the philosophy of nature, a movement which managed to be both scientific and romantic at the same time. In nature-philosophy, all organic forms were seen as external manifestations of a great inner compulsion towards perfection, with human form as its destiny. That is, the very purpose of the unfolding process was to lead, from inaccessible generation, to greater and greater perfection, with Man as the yardstick against which the forms of all creation should be measured: an expression put most forcefully by Lorenz Oken (1779–1851), who asked, 'What is the animal kingdom other than an anatomized man, the macrocosm of the microcosm?'[3]

Given their fondness for innate drives and compulsions, Wolff's *vis essentialis* found a ready audience among nature-philosophers, and it was this concentration on the ineffable that finally broke the stalemate of preformationism. After a century and a half of intense effort, preformationism was actually no closer to solving the central problem of generation than it had been in the days of Malebranche – identifying what it was that creates form from the formless. The preformationists had become so obsessed with 'generation' – concentrating on the earliest moments of life, which could not be observed with the technology available at the time – that they ignored the intricacies of embryology that followed, which could be watched unfolding in any microscope. The nature-philosophers finessed the problem by switching their focus to matters that could actually be observed.

The nature-philosophers regarded the problem of generation as intractable not only because the techniques available could not match the imagination of theorists, but also because intractability was, in their view, its very essence. Form was impressed on the formless by the same quasi-mystical force that shaped organic beings and drove them to human perfection, and that, for the nature-philosophers, was that. The *vis essentialis* of Caspar Friedrich Wolff seemed exactly such a force. To them, the study of the earliest moments of generation was as impossible, even in principle, as describing the earliest instants of the Universe or the insides of a black hole might be to a modern physicist. The more pragmatic among the nature-philosophers might have said that even though we could have no idea how generation worked, the fact was that it did, otherwise we would not be here. And once generation worked, embryos invariably and inevitably developed into adults, irrespective of whether their organs were present in pre-shrunk form and successively unwrapped, or developed in sequence by some epigenetic programme as expounded by Harvey and Wolff. To a nature-philosopher, the creation of form was a given, and it was therefore our duty to study its unfolding.

From this potent mixture of romantic and pragmatic motivations there emerged a school of German embryologists with their heads in the clouds but their feet firmly rooted in the soil of meticulous observation. From the seemingly unlikely beginnings of nature-philosophy came the modern science of embryology, developmental biology and – ultimately – the means and the desire to map the human genome.

The most prominent exponent of nature-philosophy was Johann Wolfgang von Goethe (1749–1832) – poet, dramatist, scientist and philosopher.[4] Perhaps best known as the author of *Faust*, Goethe has been called the German Shakespeare. But this would be a fair comparison only had Shakespeare taken time between plays to study the structure of the human skull, perform experiments on the nature of light and colour, and write a treatise on botany. This was all in a day's work for Goethe, who can be said to have invented comparative anatomy, who investigated the growth of plants, and who founded what came to be important natural history collections. He was one of the first to consider that there might have been an ice age, and he helped set up the first system of weather stations. His *Farbenlehre* ('Theory of Colour') of 1810 was a challenge to Newtonian optics. In terms of pure physics, it was a glorious failure. But as an expression of the importance of the qualitative and subjective in scientific research, it was a poetic victory. In Goethe we see the apotheosis of nature-philosophy as a romantic reaction to the cold analysis of Newton, seeking to place Man, once again, at the centre of things, and reinvigorate the subjective and the aesthetic in scientific observation. Given Goethe's immense productivity and wide range of interests, it is very hard to choose the single field in which he excelled. However, in the context of this book, Goethe's single most important contribution was the coinage of the term *morphology* to mean the science of form. This invention created the conceptual space in which could be discussed such issues as the relationships between different organic forms, and, indeed, the origin of form itself.

Goethe's first introduction to science came through the esoteric byways of alchemy. When recovering from a bout of overwork while a student at Leipzig in 1768, he was given the works of Paracelsus as gentle, recuperative reading. Thus started a lifelong love of science – a science in which the observer and the observed play equal roles. Goethe wanted to be able to see beneath the surface of nature to grasp its underlying unity. The processes of nature were not objective reality, on display for any independent spirit who cared to examine them, but a 'conversation' between the 'idea' of the phenomenon – the archetype – and the environment in which it was to give expression.

When watching the growth of plants, Goethe saw the 'idea' of the leaf change according to the environment, to manifest itself as real leaves – but also as petals, and sepals. This theory was touched on by Wolff, and has received some validation from modern genetics:[5] unless directed to do so by certain genetic programmes, the parts of plants that would otherwise form petals and the other parts of flowers develop as leaves. In the same way, Goethe saw the 'idea' of the vertebra, each bony segment of the backbone, moulded by reality to create the human skull, which was then thought to have been modified from several vertebrae that had been morphed together. This idea has received less validation from modern genetics, although it is now well established that in the absence of particular genetic signals, bones in the back of the skull tend to resemble vertebrae instead. To a nature-philosopher, the discovery that natural form was based on underlying templates which could be dis-covered was evidence of guidance by a divine principle. Were this not the case, the argument went, there would be no reason why natural form should not be completely random.

Furthermore, these patterns were evidence of a general striv-ing towards natural perfection, whose most perfect form was Man.[6] In support of this view, the nature-philosophers drew parallels between the increasing complexity of embryos as they grew from a unicellular state, achieving greater levels of sophisti-cation and order, and the apparent pattern seen in nature, in which animals and plants can be arranged in series, from uni-cellular protozoa, through simple polyps and so on, to fishes, amphibians, reptiles, mammals and finally Man. This scheme resonated with the theologically inspired view of the *scala naturae* or 'great chain of being', an ordering of nature according to its perceived complexity, as a way of justifying Man's position as the crown of Creation.

Despite this seemingly mystical approach, Goethe was a careful observer and deeply conscious of the need for accuracy in observation – but unlike modern scientists, schooled to produce the dry, seemingly objective prose of the technical report in which self-abnegation is the rule, Goethe spoke for subjectivity and the richness of allusion that is seen so rarely in scientific work today. The result was that even in his own time, Goethe was branded a dilettante: in his words, his scientific critics 'forgot that science arose from poetry, and did not see that when times change the two can meet again on a higher level as friends'.[7] Although modern scientists would, quite rightly, be deeply suspicious of the mystical 'essential forces' proposed by Wolff, some have begun to reappraise Goethe's view that the observer has an important part to play in the act of observation. Many would agree that scientific literature might benefit greatly from a literary, if not poetic, sensibility.

Although the mystical and somewhat overwrought language of nature-philosophy has long since disappeared from scientific discourse (it is no coincidence that nature-philosophy grew up in the context of a more general romanticism that found expression in the gothic novel as well as in poetry), the philosophy was vital to the development of biology and the quest to understand how form emerged from the formless. It was nature-philosophy that gave to scientists the concepts necessary to describe connections between embryology and evolution – fields which might otherwise have remained entirely separate. To the scientists of today, the likes of Wolff, Oken and Goethe seem almost uncomfortably idealistic, but without them it is questionable whether we would have been able to relate the development of a single embryo to the course of human evolutionary history, as I did in Chapter 1. It would, of course, be a gross exaggeration to say that without Goethe there could have been no Darwin – but had Goethe never lived, Darwin's thinking might well have been different.

Goethe died in 1832, when Darwin was a year into his voyage aboard the *Beagle*. Darwin set sail amid a storm of revolution in Europe. It was in the early months of 1830 that the citizens of France began to chafe under the increasingly authoritarian rule of their monarch, Charles X. On 2 August, the King abdicated. On that very same day, one Frédéric Soret of Geneva happened to be in the German city of Weimar, where he looked up his old friend, Goethe. Everywhere the talk was of politics and revolution, and as Soret entered Goethe's house, the aged poet asked his visitor's opinion of the momentous events of the day. Soret, thinking that Goethe was talking of the unrest in France, made an appropriate comment about the fall of kings. Goethe remarked that they were talking at cross-purposes – his mind was not on the political situation, but roaming an altogether larger canvas: on a debate between two titans of French science, which after simmering for decades had erupted at the Académie des Sciences in Paris.

The circumstances of the debate, between Etienne Geoffroy Saint-Hilaire (1772–1844) and Georges Cuvier (1769–1832), was as dramatic and bipolar as any nature-philosopher could have wished for. On the face of it the argument was over a dry, academic point – whether the form of an organism was dictated more by its morphological affinity in the great scheme of things, or by its present utility. Cuvier saw in each organic form a creature adapted to its present function. The wings of bats and the flippers of whales spoke to their adaptation, respectively, for life in the air and in the oceans. Geoffroy, in contrast, looked for underlying themes that would unite superficially very different structures, themes that spoke to an order in nature deeper than mere present function. Bats and whales may look very different on the surface, but their skeletons show clear evidence of having been made to a common pattern. Both men were partly right. The debate might have remained a footnote in obscure histories of the state of biological thought in the early nineteenth century, were it not for the personal circumstances surrounding the argument. Geoffroy and Cuvier were great scientists, whose mutual antipathy was piqued by their having once been great friends and colleagues. Their early comradeship had perished in decades of rivalry.[8]

In 1788, a year before the Revolution, the teenage Geoffroy found himself in Paris, ostensibly to study for the priesthood. He was seventh in a family of fourteen children of a French provincial barrister, so the priesthood was virtually his only career option. But religion, sadly, was not to his liking, so he decided to follow in his father's footsteps and turn to the law instead. Too late, he found that the legal profession excited him no more than had divinity, yet this dutiful son plodded on and got his law degree in 1790. In the meantime, and with the noise of the Revolution all around him, Geoffroy discovered his true love – science – and within that, the study of minerals. That was when the outside world finally impinged on the young man, and thanks to a whirlwind of unlikely events sparked by the Revolution, he woke up one morning to find himself professor of zoology at a brand new institution, the Museum of Natural History. It was 1793; he was just twenty-one, and of zoology he knew nothing whatsoever. Yet Geoffroy's legacy to zoology and the world was to prove profound and long-lasting.

Geoffroy had hardly found his bearings when he made contact with someone of his own kind – another precocious talent from the provinces. Georges Cuvier was then a young naturalist earning a living as a private tutor in Normandy. Cuvier was born in a French-speaking part of what was the otherwise German Duchy of Württemburg, which became part of France in 1793. Geoffroy brought Cuvier to Paris. The two young men could hardly have been more different: Geoffroy, from a solid French, bourgeois, Catholic background, secure foundations on which he was to build romantic castles of wild, speculative natural history; Cuvier, a Protestant in Catholic Paris, from a disputed borderland, for whom any theories and ideas would have to be built with the utmost care for him to feel secure. At first the two men were firm friends, dining together, even living together. By 1830 they had become confirmed enemies. On the eve of another revolution, Geoffroy and Cuvier locked horns in a public debate about whether form or function were most important in explaining the shapes and habits of organisms.

Both men were concerned with discovering underlying rules or 'laws' that explained what seemed, to them, to be the unity of nature. Both were keenly aware of how similar animals were beneath the skin. When you remove the superficialities of shape and colour, of size and behaviour, from organisms – especially vertebrates – you find that they are constructed along much the same lines. To return to an example I gave earlier, bats look as different from whales as you can imagine. Yet both are vertebrates, with bony skeletons. Although the arms of bats are small, light, spindly and support membranous wings, and the arms of whales form stout paddles, both are built according to the same underlying plan. In each case a shoulder girdle sprouts an upper-arm bone, or humerus, which articulates with the shoulder by a ball-and-socket joint. At the elbow, the humerus meets the radius and the ulna, the paired bones of the forearm. These bones meet the various wrist and hand bones, and terminate in five digits. Such similarities, reasoned both Cuvier and Geoffroy, were evidence of fundamental laws of form. God's plan was not random: it had rules, and it was the task of the naturalist to discover what those rules were.

Cuvier looked at the animal world and saw four fundamental groups, which he termed *embranchements* – the Articulates (insects and other jointed-legged animals), Molluscs (including clams, squid and other molluscs, as well as sea squirts, and various worms), Vertebrates (animals with backbones, including humans) and Radiates (starfishes, jellyfishes and other radially symmetrical or 'wheel'-shaped animals). A member of one of these groups was fundamentally different from a member of any of the other three, so that no comparison between them was possible. Variation within each group was a product of function and design. Bats and whales, for example, both had the basic structure of a vertebrate, but each was marvellously suited to its habitat, whether in the air or in the ocean. Crucially, animals were not collections of disassociated bones, but integrated structures. If the arm of a bat was an adaptation to life in the air, then every other part of the bat must be an adaptation to the same habitat. Adaptation in one structure would presuppose adaptations in others. When Cuvier excavated fossil creatures from the Paris region and raised, for the first time, the possibility that some of God's creations might have become extinct, he also took pride in being able to reconstruct the appearance of the entire animal from rather small pieces of bone. Such, he claimed, was possible only if one understood the importance of functional integration, of the interdependence of parts.

As a child of political uncertainty, whose early life had been subject to the whims of revolution and discord, Cuvier had become a cautious conservative, as well as exceedingly ambitious. His talent, once coupled with an unstoppable drive for authority, soon made him the most important figure in the French scientific establishment. Promotion was impossible without his patronage. Given his position, Cuvier reserved for special disdain any theories welcoming of subversive ideas such as 'transformation' (what we'd now call 'evolution'), as personified by Jean Baptiste de Lamarck (1744–1829), the elder statesman of French natural history and someone who should have been Cuvier's academic superior. In his *Philosophie zoologique* (1809), Lamarck had developed a sophisticated theory of evolution in which organisms retained a *besoin* or 'need' to improve themselves, as if they were climbing a ladder from a simple state to one more exalted and complex. Lamarck's idea bears some resemblance to nature-philosophy (the *besoin* being equivalent to a *vis essentialis*) but with one important difference. In nature-philosophy, inner compulsions towards perfection were ideal or metaphorical, and did not imply that lineages of organisms actually transformed or transmuted themselves into different species over time. In Lamarck's view, that is exactly what was meant to have happened – more complex creatures, such as Man, evolved from simpler ones. *Philosophie zoologique* contained the most coherent theory of evolution to emerge before Darwin's, precisely half a century later. Ideas of transformation were complete anathema to Cuvier, who saw them as threatening manifestations of radicalism and sedition. When Lamarck died, Cuvier gave the official eulogy – as his position, at the height of the French scientific establishment, demanded. His assessment of Lamarck was one of belittlement and barely concealed scorn.

Geoffroy, on the other hand, was given to artless forays into the intellectual wilderness. He was ever prone to grandiose speculations on the connectedness of all life, and even on the nature of the Universe – speculations that his friends did their best to rein in, fearful of the detrimental effect that such musings, were they made public, would have on Geoffroy's career. Their efforts were not entirely successful: Geoffroy's eccentric outpourings ensured that he remained on the sidelines, to be outflanked and outranked by Cuvier, his former protégé. Although he outlived Cuvier, the aged Geoffroy came to be regarded and indulged as a crank.

Geoffroy was not concerned with function, except inasmuch as it was, in his view, subservient to form. To him, the adaptations of the arms of bats and whales were as superficial a consideration as their size or colour. The remarkable fact that they were both constructed in the same way was far more important. Indeed, this fact spoke against the primacy of function. Consider: if bats were created by God to be flying animals, and whales to roam the oceans, why should they show any similarity of form whatsoever? The very fact of form must be a sign of God's purpose. And so Geoffroy searched for such signs of similarity in the animal kingdom – even when their discovery required extraordinary feats of mental gymnastics.

There was much interest at the time in the origin of series of repeated structures, and the degree to which organic forms in general were the products of repeated units, even if this origin was hard to discern or entirely hidden. Repeated or 'serial' structures can be found everywhere in nature, from the segments of insects to the vertebrae in the spine and the petals of flowers. For example, there had been much argument about whether the human skull, with its complexity of interleaved bones, could be considered as having been derived from a number of separate vertebrae. As I noted above, Goethe expended much effort on this point. Nobody then really doubted that the skull was derived from vertebrae: the argument was about the *number* of vertebrae that formed the skull, and estimates varied from three to six.[9] Geoffroy joined this argument, but as was typical, he extended it far beyond previous limits. He suggested that not only could the skull be seen as a derivative of vertebrae, but so could many other structures. He even extended the concept to far-flung members of the animal kingdom. Crossing the boundaries of Cuvier's *embranchements*, Geoffroy supposed that the segments of insect bodies followed the same formal construction as vertebrae. Insects, being hollow with an external skeleton, were like vertebrae, constructed as hollow cylinders around the space occupied by the spinal cord. Insects were living inside their own vertebrae.

To make matters worse, Geoffroy was hardly consistent in his views. His mind ranged across all human knowledge, gathering bizarre connections in the way that a jackdaw collects sparkly trinkets. One of his ideas was that vertebrates were the 'same' as insects, only turned upside down. Vertebrates have a nerve cord that runs along the back – the spinal cord – whereas the heart, aorta and other major vessels of the circulatory system run along the belly. The reverse is true in insects and other articulates. Just turn a lobster upside down, and *voilà* – a vertebrate! Cuvier and his followers regarded such schemes as preposterous – the nervous and circulatory systems in each *embranchement* would be true to their own selves, but would not be comparable to those in other *embranchements*.[10]

Geoffroy was marginalized, but he refused to go quietly. Rather than fading away in middle age, Geoffroy found his views adopted by a new generation of idealistic young naturalists enthused by nature-philosophy. This is not surprising, given that nature-philosophy strove to find the single plan on which the whole of nature was based, and if that quest required bold leaps of the imagination, then so much the better. Suddenly, and no doubt to his amusement, Geoffroy saw his ideas come back in to fashion. No longer dismissed as an ageing eccentric, he found himself the idol of a subversive counter-culture. The renewed enthusiasm for Geoffroy's ideas was a small part of an increasingly heated political atmosphere in which people began to chafe against what they saw as the heavy-handed establishment culture of the restored French monarchy. Cuvier, now the arch-conservative, came under threat both politically and scientifically as the tide began to turn against him.

In the fractious atmosphere of the time, the differences between Cuvier and Geoffroy first festered and then broke out into the public arena, culminating in a series of blasts and counterblasts before the Académie des Sciences, the premier scientific organization in France, in the spring of 1830. It was Geoffroy who struck the first blow.

The immediate cause was a paper entitled 'Some considerations on the organization of molluscs', submitted for consideration by the Académie in October 1829, by two obscure naturalists called Meyranx and Laurencet. They had worked out that if you take a cuttlefish, as a representative mollusc, and treat it to a thorough session of theoretical origami, it would resemble a vertebrate in its overall organization. Conversely, if you bent a vertebrate over backwards, so that the back of its head touched the base of its spine, it would resemble a mollusc, as regards the relative positions of its organs.

Such speculations were music to Geoffroy's ears. The original paper no longer exists, and it is possible that Geoffroy exaggerated and distorted the paper's conclusions when he prepared his report on it for the Académie on 15 February 1830. Irritated that Geoffroy had not quietly disappeared, but had found yet more ammunition for his outrageous views, Cuvier rose to the challenge, and the two giants of French biology sparred before the Académie until, on Geoffroy's initiative, the debate ended – predictably without much conclusion – in April. It was as classic a clash as any scriptwriter could have wished. These potent ingredients attracted considerable press attention, even as the political atmosphere grew more charged towards the outbreak of revolution.

Cuvier died two years later. Geoffroy survived until 1844,

but his moment of glory had passed and he disappeared into the fringes once more. The echoes of the debate, however, resounded long after silence descended on the main protagonists. To observers such as Goethe, the Académie debate was of absorbing interest. If nature-philosophy can be described in a sentence, it was the search for the origin and disposition of natural form. Understanding the origins of form had united the young Geoffroy and Cuvier in a shared passion to bring reason to the world of nature, but the quest had finally driven them apart. The Académie debate shaped the intellectual landscape for decades after 1830, and doubtless affected the first scientific thoughts of an impressionable young naturalist who set sail on a voyage of discovery the very next year. That naturalist was Charles Darwin.

5

Evolution

For us today, the connection between individual development and evolution is something that can be taken for granted. Back in the eighteenth century, however, there would have been no reason to expect or imagine that the embryonic development of an individual should have anything to do with the overall plan of nature. We have nature-philosophy to thank for making this connection explicit, for it sought correspondences between the microcosm of individual development and the macrocosm of nature.

The science of comparative embryology that sprang directly from nature-philosophy in the mid-nineteenth century shed the mystical baggage of its progenitor, but still held strongly to the notion that the events in the development of the individual reflected its station in nature. Without knowing they had done so, the nature-philosophers had managed to explain, in one coherent scheme, two essential properties of the genome: first, the course of individual development, and second, how changes in development are recorded in evolution, allowing the careful experimenter to use embryology to unlock the evolutionary history of a species.

Nature-philosophy would never have been able to explain another property of the genome – the origin of diversity. Nature-philosophers still saw the world as a static statement of Creation in which each species represented an instance of some divine plan, or archetype. If nature was to be anatomized Man, then the reason lay in the divine plan, and that was that. The abundance of one species or another was not something that could be explained without recourse to the divine. Had God intended there to be a million species of beetle but only one coelacanth, then that choice rested solely with God, and it would be as pointless to ask why matters should not be otherwise as it would to speculate on the metaphysics of beings that might exist on other worlds.

There was another reason why the conception of a static nature militated against any investigation of diversity. An understanding of diversity rests on an appreciation of variation – the small differences between individual animals or plants. Nature-philosophy, though, stressed similarity. In their constant striving to find patterns to unify disparate nature, Goethe and Geoffroy would have regarded natural variation as a distraction. The discounting of variation was a consequence of the adherence of nature-philosophers to the classical ideal of the archetype, of which each earthly instance was but a poor copy. The fact that mundane cats may be black, white or tortoiseshell was therefore never of interest in itself, just a consequence of earthly imperfection – it would never be possible to achieve an exact representation of the divine conception of 'cat' in an imperfect world. So matters stood until Darwin's insight that variation was not, in fact, the noise that cluttered up the signal. Variation *was* the signal.

The young Charles Darwin (1809–82)[1] was the unpromising scion of a wealthy doctor's family in the English Midlands, related to the famous Wedgwood ceramics dynasty. Charles's grandfather, Erasmus Darwin (1731–1802), was a noted scholar in what might be called an English strand of nature-philosophy. Erasmus was fortunate in that he lived in an age of cultural enlightenment and religious tolerance. Expansive in figure and dress as well as prose and poetry, he was able to entertain ideas of the transmutation of species, and was thought only mildly eccentric for doing so.

The Darwins and the Wedgwoods were nonconformists – that is, they were Christians, but unattached to the established Church of England. Erasmus's grandson, though, grew up during a period of increasing political and religious conformity in England, a reaction to the revolutions that had struck Europe in the late eighteenth and early nineteenth centuries – the same revolutions that had shaped the ideas and careers of Geoffroy and Cuvier. Not that such elevated themes preoccupied the young Darwin. He preferred robust country pursuits to scholarship – he was a hearty, outdoorsy type much given to hunting and shooting, with few cares about how he might make his way in the world. In 1825, aged just sixteen, his respectable, physician father sent him to Edinburgh to study medicine, but Darwin found the anatomy demonstrations stomach-churningly grisly. While he was there, however, he fell under the influence of lecturers on the radical fringe of academic society who were prepared to entertain the idea that species were not fixed by divine law, but could change according to the circumstances in which they found themselves. This idea of the transmutation of species was, by the 1820s, considered as seditious as anything in revolutionary Europe – something to be scorned by establishment scientists such as Cuvier – whereas in Erasmus's day they were thought merely extravagant and exotic.

After a year of medicine at Edinburgh, Darwin drifted into natural history before dropping out completely. The standard destination for people in his position in those days was the established Church of England, and so the squeamish Darwin was dispatched by his despairing parent to Cambridge, where in due course the young man took a degree in divinity. At the time, the old universities – Oxford and Cambridge – were seats of Anglican conformity. College fellows, irrespective of what they taught, were first required to be ministers. Darwin received his instruction in botany from one such divine, the Reverend John Stevens Henslow (1796–1861), who became a firm friend. In an alternate universe Darwin would have become a minister, taken up a quiet country living, and spent the rest of his life botanizing and sermonizing. But his life was about to take a most unexpected course.

The early nineteenth century was the great era of maritime exploration, particularly in Britain, then at the zenith of its naval might and political influence. Ships were dispatched to remote places to make observations for accurate maps and to record anything else of interest – including the indigenous flora, fauna and geology. One such survey ship, HMS *Beagle*, was due to set sail on a trip around the world, concentrating on a survey of the Atlantic coast of southern South America.

The long isolation of such voyages posed a problem for the *Beagle*'s captain, Robert Fitzroy (1805–65). Social niceties prevented him from fraternizing with his crew – yet the effective solitary confinement this imposed would strain his mental health. Fitzroy was well aware of the insanity that ran in his family. A relative, Robert Stewart, Viscount Castlereagh, a politician of consummate skill who had served as Foreign Secretary throughout the Napoleonic Wars, had committed suicide. What Fitzroy needed was a companion who could sustain five years of dinner-table conversation at a level sufficiently elevated to save him from his inherited demons.[2] News of this opening came to Henslow, who put Darwin's name forward. Darwin was affable and sociable, with the bearing and background of

69

a gentleman, and so might have been expected to be capable of cheering up the morose captain. Henslow, of course, recognized Darwin's acumen as an observer of nature and thought he would profit from exposure to the biological diversity of the tropics. Darwin's father, no doubt, hoped that this expedition would be the making of his feckless offspring.

On 27 December 1831, Darwin set sail on the *Beagle* as unpaid companion to the captain. In this task he was largely unsuccessful. Although they were social equals, they were poles apart politically and religiously. Fitzroy was High Church and High Tory; Darwin was a Whig (that is, a Liberal), and beneath a thin Anglican veneer there lurked the nonconformist tendencies of his Darwin ancestors and Wedgwood cousins. Darwin spent less time than planned with the captain, and more on natural history, to the chagrin of the official ship's naturalist, who abandoned the expedition when the *Beagle* reached South America. Apart from some seasickness, Darwin was in his element – collecting flora, fauna and fossils with enthusiasm and sending crates of specimens back to Henslow and other scientists, all the time making acute notes on everything he could find. After a long period surveying the coasts of Brazil and Argentina, during which time Darwin spent extended periods exploring ashore, the *Beagle* rounded Cape Horn, and in the autumn of 1835 spent a month among the bleak, rocky Galapagos Islands in the Eastern Pacific. The voyage continued across the Pacific to Australia and then, by degrees, home. Darwin made landfall in England on 2 October 1836.

The voyage of the *Beagle* had indeed made a man of Darwin. The cornucopia of specimens he had sent home over five years, not to mention his popular travel book, *The Voyage of the* Beagle (1839), had turned the college drop-out into a respected scientist and minor celebrity. Yet he found his fame uncomfortable. He hid from the public eye in a quiet Kent village, choosing the sedate life of the leisured gentleman, marrying his cousin Emma Wedgwood in 1839. He fathered ten children, performed a multitude of experiments in his home laboratory and greenhouse, and, painstakingly, wrote up his research.

There has been much speculation about why it took Darwin more than twenty years to publish the idea of evolution by natural selection. The answer is complicated. A busy family life would have been one factor: from 1839 until 1856, almost the entire period in which Darwin was collecting his thoughts on evolution, there was at least one school-aged child at home. During this period, Darwin was frequently laid low by a series of debilitating illnesses whose cause – or causes – remain mysterious. The pervading atmosphere of political conservatism and religious conformity, antithetical to radical ideas such as the transmutation of species, would have been another reason for his holding back. Darwin might have felt he owed too great a debt to his father, to Fitzroy, to his various patrons, such as Henslow, and most of all to his devout wife, to be able to publish such an idea with impunity. And there is another, more mundane reason: the idea took time to coalesce into an easily expressible form, especially as the idea of natural selection broke genuinely new ground in the history of thought. Darwin, a naturally cautious man, painstakingly exhausted every avenue of investigation before going public.

Conventional mythology has it that Darwin's visit to the Galapagos Islands was pivotal in the genesis of the idea of natural selection, but it was not as simple as that. This remote group of dry, volcanic islands is home to a range of peculiar creatures not seen anywhere else in the world. Many of the islands have (or had, until recently) their own species of giant tortoise. There are iguanas that – unusually for lizards – swim in the sea, and a species of flightless cormorant. There are also several species of finch, each unremarkable in appearance but adapted to a particular dietary niche. Some have thin, probing beaks for extracting insects from crevices within bark. Others have thick, robust beaks for cracking nuts and seeds. Darwin looked at these finches as a group and speculated that they were all descended from a single ancestor, or a few ancestors, that arrived on the islands from mainland South America, presumably by accident. Once there, each island's finch population became adapted to the local conditions. After millions of years, the process of adaptation to these different environments – natural selection – moulded the ancestors of the birds in each habitat until, generations later, the result was the variety of finches we see today. It would not be true to say, however, that Darwin observed the finches, made a few notes and had a 'eureka' moment. The visit to the Galapagos Islands made relatively little impression on Darwin at the time. Besides, there is no reason to suspect that Darwin's own thinking had advanced enough for him to know that the fauna of this out-of-the-way place was significant, apart from its evident yet parochial oddity.

It is a common mistake, when looking back at the history of figures that posterity has made into heroes, to attribute to those people an insight they did not have. Darwin was not a Darwinist – he was a conventional Christian, more or less, although one who, through his family connections and university experiences, had come up against exotic ideas such as the possibility that species might not be immutable, but could in some circumstances be transmuted into new forms. The Darwin who boarded the *Beagle* would have been sympathetic at least to this idea of transmutation, in the same general way that his family had historically entertained liberal views on politics and religion. Darwin was no out-and-out transmutationist, and even if he were, he would have had no greater insight into the possible mechanisms of transmutation than would any other well-informed observer of his day.

The exposure of this naturally observant young man to the wealth of tropical diversity left him intrigued by the possibilities of variation and transmutation, but it was not clear to him how the two might be related. It is certain that Darwin – before, during and after his voyage – was well-read. He knew and admired the work of European scholars such as Von Baer and Goethe, derived from or influenced by nature-philosophy. These scholars saw a pattern to nature, but not a pattern understood to have been generated by the transmutation of species. Nevertheless, nature-philosophy might have given Darwin a clue. Nature-philosophers stressed that the largest patterns in nature were evident in the smallest things, that the microcosm mapped the macrocosm. The early embryologists, inspired by nature-philosophy, applied this philosophy to the origin of form, showing that every embryo replays its own evolutionary heritage over the course of its development. Darwin, however, took nature-philosophy in a new direction by showing how individual variations on the most trivial scales could, through transmutation, be harbingers of the greatest evolutionary changes.

A problem for anyone sympathetic to the idea of transmutation of species was the great length of time that the process would require – far greater than the few thousand years allowed for in the Bible, then the supreme authority on the history of the Earth. If any changes had happened on the planet since its creation, they would have to have been sudden and dramatic. When fossils started to be collected in a systematic way, in the eighteenth and early nineteenth centuries, and it became evident that the creatures to whom these bones belonged were extinct, it was suggested that they perished in a sudden catastrophe. The biblical flood of Noah was said to stand for several cataclysms in which organic life perished, each time to be created anew in an improved form. Cuvier was a prominent advocate of this 'catastrophist' view.

A new strain of thought emerged among geologists who found that evidence for convenient disasters was rather thin on the ground. They suspected that catastrophes were something that characterized ineffably remote ages, and happened rarely, if at all, in the modern world – in the same way that the Deity seemed to have spent a lot of time intervening directly in the lives of the Patriarchs, but had lately got out of the habit of manifesting himself as a burning bush or turning people into pillars of salt. Examining the world around them, what geologists saw was a landscape shaped by the kind of gradual processes we see all around us today: the slow accumulation of change in which, for example, raindrops take millions of years to turn a mountain, grain by grain, into a valley, rather than the violent irruptions required by the catastrophist view, which conveniently allowed for great change to be accommodated within the chronological confines of scripture. This new tendency of geological thought, set up in opposition to catastrophism, was the uniformitarian school, and its leading exponent was Charles Lyell (1797–1875).

The first volume of Lyell's book *Principles of Geology* had been published just before Darwin embarked on the *Beagle*, and it was Darwin's favourite reading during the voyage. The tale of the slow, gradual processes that have shaped the Earth, over immensities of time not accounted for in the Bible, must have spoken to the nonconformist, liberal side of Darwin's nature. If the Earth was immensely – perhaps immeasurably – old, there would have been time enough for species to transmute, and Lyell even suggested a form of transmutation in which species would become adapted to their environments. Lyell's book provided the backdrop against which Darwin would eventually stage his drama of evolution by natural selection.

When Darwin arrived back in England, he had other matters besides transmutation to occupy his mind. There was the curation of the specimens he'd sent back from the *Beagle* voyage; there was marriage, children, writing up his adventures and his continuing and prolific researches into everything from the taxonomy of barnacles to the growth of orchids. Amid all this activity he began to arrange his thoughts about transmutation. He set out to synthesize his own observations, from the *Beagle* and from his experiments in the greenhouses and gardens at his home, Down House, with the Lyellian view of geology, with the picture of idealized nature as promoted by the nature-philosophers, and with the more revolutionary ideas of transmutation as suggested by his grandfather and his teachers at Edinburgh.

Some years elapsed before Darwin started to bridge a gap that nobody else had dreamed even existed. The great theme of transmutation over the unthinkably vast intervals of time suggested by Lyell might be rooted in the commonplace variation between individual creatures that we can observe in the here and now – a new reading of the nature-philosophic connection between the microcosm and the macrocosm. Time and chance acted on finches, on giant tortoises, on breeds of dog, to produce a distinct variety, each suited to its own circumstances as Lyell had suggested. But how would each variety become suited to its environment, rather than vary in some other way, or not at all? Darwin had an answer in a mechanism he called natural selection. This term is shorthand for the action of environmental circumstances on a population of creatures that produces more offspring than there are resources to support them. As a result, only those offspring best adapted to the prevailing circumstances will survive for long enough to reproduce themselves. Over many generations, the complexion of the population will come to reflect those better-adapted offspring and their descendants. For this scheme to work, the population has to be varied in constitution, and this variation has to be inheritable.

Crucially, natural selection contains no element of destiny or direction. It was only later that scientists such as Ernst Haeckel (1834–1919) misappropriated natural selection as a motor for a kind of evolution based on destiny – the kind of quest for perfection promoted by nature-philosophy. However, Darwin himself was not immune from such ideas – and why should he have been? In his notes on transmutation, and reading widely from the then-current nature-philosophy, he tried out a number of concepts to see how they suited, including the evolution of complex life from spontaneously generated particles, or monads; the idea that species, like individual creatures, have cycles of life and death (a notion inspired by Lyell's own ideas about cycles in geological time); and the idea that transmutation might have some directional, progressive character. But spontaneous generation, monads and ideas of progression were gradually stripped away as Darwin came to understand that only three things are necessary for natural selection to work – time, the fact that creatures generally produce more offspring than the environment can support, and the inherited variation in these offspring.

Lyell's *Principles of Geology* provided Darwin with the ingredient of time. The element of superabundance came later, after Darwin had begun to construct his theory of evolution as something directed and progressive – which was ultimately antithetical to his final theory of natural selection. Such ideas were abandoned some time after 1838, when Darwin read *Essay on the Principle of Population* by the economist Thomas Malthus (1766–1834), first published as a pamphlet in 1798 but expanded in successive editions up to 1826. Malthus's essay started with the premise that whereas populations grow in geometric progression, the resources necessary to support them will grow at a slower, arithmetic rate. Thus populations will outstrip their resources until checked by famine or war, if not by self-restraint. This idea struck a chord with Darwin, who found within it the ingredient he needed – a mechanism that would make evolution work.

Malthus was working during the Industrial Revolution, perhaps the greatest social upheaval in England since the Black Death of the fourteenth century. During the Industrial Revolution, the social, economic, political and even the physical landscape changed beyond recognition in the space of a few decades. People left the land and the populations of cities surged. Some people, such as Darwin's Wedgwood cousins, grew rich from industry. Many others remained poor – the new urban, working class. Crowded into slums, working people experienced poor health, epidemics of diseases such as cholera, and high infant mortality. Until the twentieth century – and then only in those societies equipped with adequate sanitation and pension provision – people relied on their children to look after them in their old age. But because infant mortality was high, people had as many children as possible in the knowledge that many had to be born to ensure that a few would survive long enough to become breadwinners themselves. Such problems were particularly acute in the overcrowded slums of the growing industrial cities, and the resulting poverty, and its alleviation, was what concerned Malthus. However, families on all social levels were stricken by the deaths of infants and children. Even the Darwins, comfortably wealthy as they were, lost two of their ten children in infancy. A third, their daughter Anne Elizabeth, died from scarlet fever at the age of ten, a blow from which Charles Darwin never fully recovered.

Darwin read Malthus and made a dramatic leap of intuition: what applied to the urban poor could be true of the entire world of nature. Darwin had long observed that animals and plants produce vastly more offspring than can ever be supported either by the parents or the resources available, simply so that a few would survive to reproduce themselves. You do not have to travel to the Galapagos Islands to see this principle in action. Every oak tree in the park produces thousands of acorns each year, each one the germ of a new tree. But if every acorn became a tree, we would soon be unable to move for oak forests. In the real world, most acorns rot or are eaten by animals long before they have a chance to take root.[3]

Darwin built upon Malthus's insight to observe that, of all the offspring produced by animals and plants, only those most suited to the environment would survive to reproduce – only those acorns most adept at avoiding being eaten would become oaks. This is why acorns are tough and poisonous rather than tender and delicious. Only the toughest and most poisonous of them will last long enough to grow into trees. (If fruits *are* tender and delicious, this is a ruse to attract birds or other animals who will eat the fruits and disperse the hard and resistant seeds within.) In the 'struggle for existence' as painted by Malthus, only a few offspring would be tough enough, or 'fittest', to survive. But because the world is a harsh place, a superfluity of offspring must be produced, just to be sure that a few will be capable of survival and reproduction themselves.

Lyell gave Darwin the time, and Malthus gave him the abundance. The final ingredient was variation. Variation is the keystone of Darwin's theory, for within it lies the possibility of change, and it was Darwin's connection of variation with change that gave him the idea of natural selection. If all offspring – whether of oaks or people – were exactly the same, it would hardly matter which among all the millions survived: it would be a lottery. But if offspring differ from one another, even slightly, the possibility is raised that some will just happen to carry traits favoured by the prevailing environment, and these individuals will have a greater chance of surviving to perpetuate the species, at the expense of their fellows. Over generations, the pool of traits carried by the species as a whole would be enriched in those favoured traits, and depleted in those less favoured. Given the large intervals of time offered by Lyell, the species itself would have the potential to transmute into others, as time and circumstances allowed.

After his return to England, Darwin started and maintained a voluminous correspondence with people of every kind, with the aim of understanding the extent of variation and how it could be moulded and influenced. Among his correspondents were stock breeders, nurserymen and pet fanciers – people who set store by the promotion of desirable traits in animals and plants, and the elimination of undesirable ones. The *Origin of Species* contains much of interest on this score, especially concerning pigeons. From common species of wild pigeon, fanciers have created a range of breeds not found in the wild, such as pouters, tumblers and fantails, by a long process of selection. A breeder will select those birds with most signs of the favoured trait (a more fan-like tail, for example) and allow only those birds to breed. The process is repeated, so that over many generations this trait will come to predominate. Such is the 'artificial' selection which Darwin used as an analogy for the grander process in which nature takes the place of the pigeon fancier, selecting which of the many offspring of an animal or plant, in any generation, will survive long enough to reproduce.

Pigeon fanciers, and breeders in general, can produce dramatic changes in the appearance and behaviour of animals and plants in much less than the span of a human lifetime. Given the immensities of time granted by Lyell and his colleagues, nature would have had ample opportunity to create the whole diversity of life from some primordial creature floating around in the deeps of time. And as pigeon fanciers can create several distinct varieties from a single stock of pigeons, so could nature produce different species of, say, Galapagos finch – from a single pair of progenitors. This leads to the insight that because natural selection, in any generation, must work with what it has to hand, and cannot start again from the beginning each time, the form of a species will be moulded as much by its heritage as by its current circumstances, a realization that immediately reconciles the seemingly antithetical views of Geoffroy and Cuvier. It follows from this species-on-species accumulation of heritage that the forms of embryos necessarily track the changing forms of the adults into which they will grow. The nature-philosophers saw the connection between individual development and large-scale patterns of form in nature, but Darwin was the first to explain why this connection existed.

Time, abundance and variation. Darwin needed time to understand all three factors before he was able to piece his theory together in its final form. But even when his theory was complete, he prevaricated endlessly rather than go into print – collating his notes, piling example on example, and building a compendium of evidence in support of his idea with the slowness of a Lyellian geological process. Darwin understood that his theory of evolution (he called it 'descent with modification') by means of natural selection was more than just another stab at transmutation, but a qualitative advance in thought, and one which people in general might find hard to accommodate. Events, however, overtook him. Like much unwelcome news, it arrived in the mail.

In 1855, Darwin came across a paper by Alfred Russel Wallace (1823–1913), then an unknown land surveyor turned professional collector of natural history specimens, who had worked in South America, but who was lately in the East Indies. Wallace reported that new species tend to appear in regions close to the home ranges of the species they most closely resemble. Wallace was close to producing a theory of evolution similar to

Darwin's, but he had not got there yet, and Darwin calculated that he still had time to collect his own thoughts before publication. Three years later, in June 1858, Darwin saw another paper by Wallace in which the younger man had hit the nail on the head, describing natural selection in a few pages, quite independently of Darwin. What was worse, Wallace had sent it to Darwin in person, to see what the famous naturalist and *Beagle* veteran thought of it.

Apart from discussing his emerging theory with a few scientific friends (including Lyell), Darwin had kept his thoughts on natural selection to himself. Now here was Wallace, arriving out of nowhere and about to put his name to an idea which Darwin felt was his own property. Darwin was thrown into agonies of moral indecision. By rights, the first person to publish should claim the credit, and that was likely to be Wallace. On the other hand, Darwin had been thinking about natural selection for many years – presumably, for longer than Wallace had been – and was working up a long and exhaustive treatise on the subject. The problem was compounded by the fact that Wallace, in his naïvety, had written to Darwin, completely unaware that he was an interested party.

Darwin did what many would have done in such a situation – he asked a close friend and mentor for guidance. The friend was Lyell, who arranged an honourable tie. On 1 July 1858, Lyell and another of Darwin's friends, the botanist Joseph Hooker, read Wallace's paper and a few short items from Darwin at a meeting of the Linnean Society, London's senior institution for biology. Neither of the main protagonists was there. Wallace was in the East Indies, and would not return to England until 1862. Darwin was at home, distracted by a spate of family illnesses, and took little interest in the proceedings in London; neither, as it turned out, did many other people. Summarizing the business of the year, the Secretary of the

Linnean Society reported that 1858 had been unremarkable for scientific advances.

The fireworks began when Darwin, spurred on by his friends, composed an 'abstract' of his ongoing work as a stopgap, to show that although Wallace had threatened to beat him to publication, he had been thinking along these lines for years and had amassed a large amount of evidence to support his case. That short book was, to give it its full title, *On the Origin of Species by Means of Natural Selection, Or the Preservation of Favoured Races in the Struggle for Life*, and it was published on 22 November 1859. This time the world sat up and took notice. The *Origin* was an immediate sell-out: orders for 1,500 copies outstripped the first printing of 1,250, and a second edition was hurriedly prepared. As for the reaction of the public, the historian of evolution David Hull put it thus: 'One can safely say that all hell broke loose.'[4]

Darwin's theory of evolution by natural selection has endured because it explains a great deal about how and why the natural world came to be the way it is. It shows how the interaction between inherited variation and external circumstances can mould, or adapt, species to their environments – and, if necessary, reshape species to meet changing circumstances. Natural selection provides a unified explanation for phenomena which might otherwise seem unconnected, from the nesting habits of certain moths that make their homes only inside the fruits of yucca plants, to the sexual preferences of peahens; from the profusion of species of cichlid fishes in the East African Great Lakes to, perhaps, why gentlemen prefer blondes. It also explains how several species can emerge from a single ancestor; why the pattern of life is hierarchical, like the branches of a tree; and, finally, how all the species of the Earth might have descended from some blob of protoplasm in what Darwin himself called a 'warm little pond'.[5]

One of the most attractive features of natural selection is the disarming elegance with which it explains so much by postulating so little. The mark of a sound theory is that it explains evidence with minimal recourse to extraneous assumptions. There had been other schemes to explain transmutation – and before that, generation – but they could be made to work only by positing the existence of elaborate microscopic structures, mysterious forces or essences, or divine intervention. Before Darwin, perhaps the most influential theory of transmutation had been that of Lamarck, who had proposed that animals evolved through an inherent improving force, or *besoin* ('need'). There is nothing inherently wrong with this as an explanation. However, Darwin's natural selection is better because it explains the natural world perfectly well without having to assume the presence of extra internal mechanisms. Natural selection required only three things, and all of them could be observed and measured – time, abundance and variation.

Natural selection was the first theory in which all the important properties of the genome – its direction of individual development, the relationship of this development to evolution, and the generation of diversity – were united in a single coherent scheme. Nature-philosophy had linked the first two, but its conception of a static nature blinded it to the third. In his insight that variation was important, and not merely a distraction, Darwin found a way to explain the diversity of nature in terms of the overall pattern of life.

There remained something that natural selection did not explain. Whereas Darwin understood the importance of variation, he could not account for how this variation was generated and maintained. Natural selection does exactly what its name suggests – it selects variant forms from a larger range of variation presented to it. But without some mechanism to regenerate that variation, selection would soon wear it away into bland homogeneity, and the transmutation of species would stop. Darwin could offer no credible mechanism for the generation and maintenance of variation. As Harvey and the pre-formationists had been with the question of the origin of embryonic form, Darwin was ultimately confronted with what looked like an insoluble problem: where did variation *come from*?

Darwin was acutely aware of this shortcoming, and it worried him greatly. He realized that variation must be inheritable, but at the time there was no clear idea about how this variation was organized or stored as information, nor how traits from the parents were apportioned in the offspring. It seemed clear that offspring manifested traits of both parents, a circumstance that had bothered preformationists who theorized that traits would be passed down through either the male or the female line, but not both. In general, the view of inheritance in Darwin's day was that the characters of both parents became indissolubly blended in the offspring. If such blending inheritance were the rule, then any variation present in a population would sooner or later be eroded, homogeneity would result, natural selection would have nothing to select, and evolution would stop.

A cursory glance at the variety and diversity of organisms around us instantly shows that this model of blending inherit-ance cannot be sustained. Variation must be encapsulated and passed down in some other way — but how? The logical sol-ution, Darwin reasoned, was a system in which inheritance is carried by inviolate, atomistic particles that can be passed down from one generation to another without destruction or alter-ation, and whose combination and assortment in the offspring might determine the expression of this trait or that. Such atom-istic theories of generation have a surprisingly long history. As long ago as 1520, the alchemist Paracelsus — the same who published a recipe for homunculi — wrote:

Both the man and the woman each have half a seed and
the two together make a whole seed. And note how they
come together. There is in the matrix an attractive force
(like amber or a magnet) which draws the seeds unto itself
. . . Once the will has determined, the matrix draws unto
itself the seed of the woman and the man from the humours
of the heart, the liver, the spleen, the bone, the marrow,
blood vessels, muscles, blood, and flesh, and all that is in
the body. For every part of the body has its own particular
seed. But when all these seeds come together, they are
only one seed.[6]

Alchemy aside, the leading atomistic theory of generation
came from the fertile mind of the naturalist and encyclopaedist
Georges-Louis Leclerc, Comte de Buffon (1707–88). In Buffon's hypothesis, particles from every organ of the body unite
and crystallize in the sperm or egg, interlocking to produce the
'essence' of a whole organism that would provide the germ for
the next generation. Birth defects would result from damaged
or missing particles, or from an imperfect melding of particles.
The physicist Pierre-Louis Moreau de Maupertuis (1698–1759)
had a similar idea that involved the attraction between oppo-
sitely charged particles, drawn from a large pool of such particles,
each of which would represent a different part of the organism.
Elaborate speculations on the hypothetical behaviour of invisible
particles were easily ridiculed by hard-nosed, experimentally
minded preformationists such as Spallanzani, who famously
remarked that 'we descend from the observations of Leeuwenhoek to those of Buffon'.[7] Neither have Buffon's views been
in any danger of contemporary rehabilitation: in *Early Theories
of Sexual Generation* (1930), F. J. Cole wrote with characteristic
asperity that:

The commanding position which Buffon occupied in the biological world in the middle of the eighteenth century, which he owed rather to an eloquent and forceful personality than to the possession of great scientific merit, was nevertheless inadequate to ensure the acceptance of his elaborate system of pangenesis, which was universally applauded but quietly shelved.[8]

Faced with no credible alternative, Darwin came up with an atomistic theory of generation which owed something to Buffon – and which met the same ignominious fate. In 1868 Darwin described an idea he called pangenesis in which each part of the body produces a microscopic representative, or 'gemmule'. The gemmules would travel to the sperm or egg cells, and the characteristics inherited by the offspring would then depend on the number and nature of the gemmules that happened to be in the sex cells at the time. Darwin's cousin Francis Galton (1822–1911) suggested that pangenesis might be tested by looking for alterations in the distributions of traits in the offspring of animals that had received blood transfusions from animals with different traits. If gemmules were carried in the blood, the offspring might resemble the blood donor as well as the parent. These experiments were carried out on rabbits and revealed no such effect. Darwin's response was uncharacteristically evasive: he suggested that gemmules must be carried by some means not involving the blood.

This, then, is the agonizing situation in which Darwin found himself: for his theory of natural selection to work, it was necessary to posit a complicated scheme of particulate inheritance for which no concrete evidence existed whatsoever. This did not, initially, affect the wide prominence enjoyed by natural selection after 1859, even if it was not universally accepted. To begin with, critics were less concerned with its failure to explain the sources and nature of variation on which the theory rested, than with the challenge that evolution posed to established religious and social views. People were quick to grasp the idea that change is the usual state of nature, shaped by external circumstances, and that organisms were not tied to divinely ordained stations with respect to their fellows. It is not surprising, therefore, that 'Darwinism' was adopted by social reformers and radical politicians, and fought with vigour by representatives of the establishment. This, of course, was just the kind of reaction that Darwin had been afraid of, and one reason why he kept quiet about it for a long time. The scientific reaction was slightly different. Unlike the social reformers who saw in Darwinism the unlimited possibilities of change offered by freedom from the divine plan, scientists saw natural selection as a progressive force for inexorable evolutionary improvement.

Blame for this perversion of evolution as an instrument of destiny may be laid, in part, at the door of Ernst Haeckel – prolific biologist, embryologist and populist of the theory of evolution. That Haeckel was predisposed to see evolution as progressive is perhaps no surprise, schooled as he was in the traditions of German embryology and anatomy that had, in turn, received their impetus from nature-philosophy. Haeckel revered Goethe as much as he idolized Darwin: he fused Darwinism with the nature-philosophic picture of the world as a series of stages striving towards the human ideal – with natural selection miscast as the motor. Haeckel fused Darwin with Goethe and created a monster. It is no accident that this bastardized Darwinism was misappropriated by the Nazis, and then, all of a sudden, otherwise quite ordinary citizens were committing unspeakable crimes in the name of the evolution of a more perfect human species. Nazism may have been defeated, but Haeckel's dismal legacy lives on. In the public mind, 'evolution' means progression and improvement. Everyone has seen images featuring a line of figures from apes to Man – sometimes culminating with the latest consumer product, accompanied by a slogan such as 'Move Up to the Next Stage in Evolution'.[9]

Natural selection is neither progressive nor a force; it has neither memory nor foresight, and works only in the here and now. Neither is it a spirit or essence, separate from the materials on which it acts. Yet it remains all to easy to adopt a view of evolution driven by an inherent bias towards improvement, and with it a kind of teleology in which structures were seen to have evolved to fulfil some pre-existing purpose. An example that has been exercising scientists recently is that of the feathers of birds, structures that seem perfectly adapted for flight. And so they are: but it is one thing to say that feathers are adapted for flight now, and quite another to assert that flight is the purpose for which feathers originally evolved. That would be to conflate the state of feathers in the here and now with the entirely separate process whereby feathers achieved that state over millions of years. Thinking along these lines, some ornithologists have asserted that feathers and flight are features that *define* birds – but this view has recently been challenged by the discovery of fossils of dinosaurs which, in life, were unlikely ever to have flown yet had structures all but indistinguishable from the feathers of birds. If feathers evolved for any purpose, that purpose was not flight – at least, not initially. To say that if all birds have feathers, then all animals with feathers must be birds is as logical as saying that if all giraffes have four legs, then all four-legged animals must be giraffes.

The trap is to link the evolution of major groups (such as birds) with the supposed adaptation of some evolving structure (such as feathers) to suit some *purpose* (such as flight). Natural selection is universally proposed as the force that drives this process, as if it were some magical ingredient akin to Wolff's *vis essentialis* or Lamarck's *besoin* that drove giraffes to grow ever longer necks so that they could pluck leaves from ever higher branches. In this way, natural selection is imbued with a memory and a foresight that it patently lacks, and evolution is reduced to the level of fairy tales. Of course, stories about the evolution of a giraffe's neck, or the feathers of a bird, are a long way from the nature-philosophic belief in the striving of organisms towards the exalted human state, but beneath all such scenarios lies a general tenor of 'improvement'.[10]

Darwin's failure to account for the origin of variation left his theory vulnerable to adulteration by older ideas of progression and improvement. As a result, the enthusiasm with which Darwinism was greeted among scientists in the 1860s had, by the 1880s, turned to disenchantment. Some scientists reverted to Lamarckism.[11] Others turned to experimental biology after the disheartening realization that *ad hoc* theories about adaptation were very largely valueless as explanations for evolutionary change. Among this latter group was William Bateson (1861–1926), who in 1894 wrote the following excoriating lines:

> Any one who has had to do such work must have felt the same thing. In these discussions we are continually stopped by such phrases as 'if such and such a variation then took place and was favourable', or, 'we may easily suppose circumstances in which such and such a variation if it occurred might be beneficial', and the like . . . 'If', say we with much circumlocution, 'the course of Nature followed the lines we have suggested, then, in short, it did.' That is the sum of our argument.[12]

These lines are from Bateson's book *Materials for the Study of Variation*, in which the problem of the origin of variation was

confronted head on. Six years later, Bateson invented a word
for the new science of variation that he and others were pioneering. That word was 'genetics'.

6

Monsters

Unlike Darwin, Geoffroy or Spallanzani, there was never any doubt about William Bateson's career. Academia beckoned from the start: from Rugby School he went to St John's College, Cambridge (where his father was Master) and took a double first in natural sciences. After a spell at Johns Hopkins University in Baltimore, he returned to Cambridge, but left again in 1910 to become the first director of the John Innes Institute in Norwich, a pre-eminent centre for the new science of genetics.

Early on, Bateson acquired the self-confidence of one who is convinced that he is in the right and everyone else is wrong. 'He formed definite opinions on a number of subjects, from the Sistine Madonna and compulsory Greek to nationalism and natural selection,' wrote the biologist J. B. S. Haldane in a review of Bateson's posthumously collected essays and addresses.[1] By the same token, Bateson was not always ready to accommodate the views of others, something his peers found a 'source of regret', according to Bateson's American colleague Thomas Hunt Morgan.[2]

Like many Victorian zoologists in the decades immediately after the *Origin*, Bateson was interested in tracing the early evolution of the vertebrates – the backboned animals, including ourselves – by comparison with our invertebrate relatives such as sea squirts and lancelets. Bateson was particularly interested in acorn worms, obscure residents of muddy beaches and ocean bottoms. These creatures, hardly known outside a small coterie of zoologists, are rather unpleasant-looking, flaccid creatures that range from a couple of centimetres to a hundred times that length. Despite their unprepossessing exterior, they have long fascinated zoologists because they seem to bridge a gap between two very different animal groups. They have a series of gill slits reminiscent of those of fishes, but as microscopic larvae floating around in the sea, before adopting the settled life of the adult, they look very like the larvae of echinoderms, the group of animals that includes starfishes and sea cucumbers.

In the 1880s, when Bateson began his work in this field, the origin of vertebrates was a hot topic, and the study of transitional creatures such as acorn worms was seen as a way to find out about a period of evolution now long vanished and yet crucial to the understanding of our own place in nature. Of particular interest to Bateson – as it remains to scientists now – was understanding the evolution of those features that make vertebrates distinctive, such as the head and the serial arrangement of muscles which allows rapid swimming. The superficially fish-like lancelet, being most closely similar to vertebrates, was the focus of special attention. Lancelets have segmental muscles arranged in a nose-to-tail series, just like fishes. Tunicates, in contrast, generally spend their adult lives rooted to the same spot, filtering particles of food from sea water strained through their gill slits. Tunicate larvae are more mobile, each propelled by a tail supported by a notochord, which are both lost when the larva settles and metamorphoses into an adult.

Several scientists looked at these facts and wove them into elaborate evolutionary scenarios. Between the 1890s and the

1920s, an English marine biologist called Walter Garstang (1868–1949), who in his regular occupation had the more mundane task of monitoring fisheries, suggested that long ago some tunicates forwent their adult stage and remained as tadpole-like larvae throughout life, gradually evolving into vertebrates. This idea was elaborated by others, notably the English tunicate specialist N. J. Berrill (1903–96), as late as the 1950s. According to Berrill, these evolving animals acquired serial muscle blocks as adaptations for efficient swimming, especially against the current. Life originated in the sea, but it was supposed that vertebrates had evolved in freshwater. On the basis of this assumption (now known to be flawed), Berrill argued that the immediate ancestors of vertebrates must have swum against the current to find fresh water – where they would have needed to evolve kidneys as a way of excreting the water absorbed by their relatively salt-rich tissues. Kidneys are organs unique to vertebrates, not found in lancelets. Skulls and brains came later, as adaptations for a new role as predators – and behold, vertebrates appeared. Lancelets, however, remained at sea – evolutionary laggards unsuited for energetic migration to freshwater. This is less a scientific hypothesis than a latter-day fairy story that brings in natural selection where needed, as a magical ingredient which transforms one animal into another, as surely as a magic spell can turn a frog into a prince, and leave a toad, or a lancelet, untouched.

In 1909, the Linnean Society staged a debate on the origin of vertebrates to commemorate the golden jubilee of the publication of the *Origin of Species* – which had first emerged in the papers of Wallace and Darwin, read to the Society in 1858.[3] On one side was a physiologist named Walter Gaskell and his colleagues, who held that vertebrates were, in essence, highly modified crabs. Opposing them were arrayed the great zoological names of the day, who held that lancelets were the closest relatives of vertebrates (as is the consensus today). One of the participants, Thomas Stebbing – a cleric, and perhaps a neutral voice – remarked that:

> When we return home and our friends gleefully enquire, 'What then has been decided as to the Origin of Vertebrates?', so far we seem to have no reply ready, except that the disputants agreed on one single point, namely, that their opponents were all in the wrong.[4]

The debate at the Linnean was the last echo of what, by then, was a long-outmoded style of evolutionary speculation.

By that time, Bateson had abandoned the quest for the origin of vertebrates as so much vacuous storytelling, but the seeds of his dissatisfaction were sown early. During visits to America in 1884 and 1885 to collect specimens of acorn worms, he had met William Keith Brooks (1848–1908) of Johns Hopkins University, a passionate evolutionist – but one who also liked to ask the kind of difficult questions about evolution that Bateson's colleagues at home seemed content to avoid. Brooks impressed on Bateson the importance of the 'species question' in biology. Darwin says that new species evolve from old ones – but how? Is natural selection the answer? And what, anyway, *is* a species? What is the nature of the variation on which selection acts? Brooks was interested in the possibility that species did not transmute gradually, but all of a sudden, in a series of jumps. Bateson was gripped by this idea of evolution through sudden mutation rather than the 'insensible gradations' proposed by Darwin, and following his meeting with Brooks his career took a new turn.

By 1894, Bateson's views on the search for the origin of vertebrates would have made a fine review of the Linnean debate of 1909, for all that they came fifteen years ahead of the event. 'Were we all agreed in our assumptions and as to the canons of interpretation, there might be some excuse, but we are not agreed,' he writes, in the introduction to *Materials for the Study of Variation*:

> Out of the same facts of anatomy and development men of equal ability and repute have brought the most opposite conclusions. To take for instance the question of the ancestry of the Chordata, the problem on which I was myself engaged, even if we neglect fanciful suggestions, there remain two wholly incompatible views as to the lines of Vertebrate descent, each well supported and upheld by many. From the same facts opposite conclusions are drawn.[5]

Bateson then drew a line in the sand, ending his career as an evolutionary biologist: 'Facts of the same kind will take us no further. The issue turns not on the facts but on the assumptions. Surely we can do better than this. Need we waste more effort in these vain and sophistical disputes?'[6] With these lines, Bateson went straight to the heart of the problem: if scientists were to make any progress towards understanding evolution, they had to try to make sense of the nature and sources of the variation on which selection acts. Natural selection may or may not be fine as a mechanism, but with no comprehension of the sources of variation we would come no nearer to understanding the nature of that which creates and maintains an organism – what we now know as the genome.

It is important to remember that Bateson wrote *Materials* before anyone had any clear idea of the mechanisms of inheritance, still less of genes or genomes. He knew that he had to start at the beginning, with a comprehensive catalogue of all instances of natural variation he could find, from which he hoped that any underlying laws of variation might emerge. To this end he scoured libraries, searched in museums at home and abroad, and assembled his own collections from as far away as Central Asia and Siberia. The resulting work – *Materials for the Study of Variation* – is not so much a book as an encyclopedia. It is a hefty compendium of no fewer than 886 reported deviations from the norm: variations in the numbers of teeth, horns, scales, bristles, body segments, sex organs, legs, antennae, digits, coloration and tentacles in a vast range of organisms, from people to pythons, shrimps to sloths.

Bateson was especially fascinated by the same patterns that had intrigued Geoffroy and Goethe – in particular in the repetition of parts, whether leaves on a stem, teeth in a jaw, or segments in the body of an insect, or a fish – and how individual elements in a series of repeated elements may be specialized. He used the term 'meristic' to refer to this kind of serial repetition. An example might be the legs of a segmented animal such as a crustacean – a lobster or shrimp. Most segments of the body of a crustacean bear paired limbs. In one sense, the limbs are 'the same' in that they are all repeated units which have much in common as regards their position (paired, attached to body segments) and structure (division into a set number of articulating parts). In another sense, each limb is specialized according to its function. Limbs associated with the head may have been turned into claws for grasping prey, or jaws for processing the prey, once caught; limbs farther back are used as walking legs; the limbs at the end are paddles, for swimming.

But Bateson also discussed disturbances in such ordered series. He described as an 'extraordinary Discontinuity of Variation'[7] those cases in which one element in a repeated series may be supplanted by that of another with a brazen neatness in its substitution. The vertebrae in the human spinal column provided several good examples. They are not identical repeated units, but the column as a whole is clearly divisible into several regions, the vertebrae in each region having a distinctive form. Like the legs of lobsters, the vertebrae in the human spine all clearly belong to a repeated series, but are also specialized according to their region. The skull is supported by the highly distinctive atlas and axis vertebrae, at the top of a series of cervical (neck) vertebrae. These vertebrae do not normally bear ribs, unlike the vertebrae in the next, thoracic region. Below the thoracic vertebrae come the stout, non-rib-bearing lumbar vertebrae of the lower back; the sacral vertebrae, fused into a single bone, the sacrum, to which the pelvis and lower limbs are attached; and finally the vertebrae forming the coccyx, the tiny vestige of a tail. Bateson catalogued instances, gleaned from anatomy, in which the lowermost cervical vertebrae tend, unusually, to bear ribs so that they resemble the uppermost vertebrae in the next (thoracic) region. In other cases, the lowest thoracic vertebra is uncharacteristically free of ribs and so resembles, instead, the uppermost lumbar vertebra – and vice versa.

More strikingly, Bateson recorded instances in insects and other arthropods (jointed-limbed animals) in which one kind of appendage appears in place of another. His case no. 76, for example, is a specimen of the bee *Bombus variabilis* 'taken beside the hedge of a park in Munich, having the left antenna partially developed as a foot'.[8] Case no. 78, in contrast, is a moth, *Zygaena filipendulae*, which possesses a 'supernumerary wing arising in such a position as to suggest it replaced a leg'.[9] Bateson coined a word, homeosis, for the situation in which one kind of appendage or member in a series (such as an insect leg, or a human thoracic vertebra) appears in the place usually occupied by another (such as an insect antenna, or a human cervical vertebra). He could hardly have predicted the importance that homeosis would assume, a century later, in our understanding of the origin of form.

Bateson's fascination with meristic variation – *Materials* is dominated by it – speaks to a deeper concern, and suggests a fundamental difference in approach from Darwin's. Whereas Darwin travelled the world and saw abundant diversity crying out for explanation, Bateson catalogued the minutiae of museum collections and saw not diversity, but unity. Parts may vary in a series, but we would not recognize the series at all without some kind of underlying similarity.

Conventional wisdom has it that 1859, the year in which the *Origin of Species* was first published, is a kind of biological Year Zero, such that discussion of work before that date becomes more of historical than of scientific interest. If this was the view in 1894, it is not reflected in *Materials*: Bateson's preoccupations with unity and diversity clearly owe much to pre-Darwinian ideas – ideas which, in the absence (to him) of any credible Darwinian consensus, were still vital and relevant. Nature-philosophy was, of course, built on the same preoccupations: Goethe, for example, had written on such things as the underlying similarities of the organs of plants, suggesting that structures as diverse as petals, sepals and stamens could be thought of as modified leaves. Echoes of the debate between Geoffroy and Cuvier can clearly be heard in Bateson's discussions of monstrosity, of vertebrae and of the structure of insects.

However, *Materials* plumbs even deeper seams of thought than these, showing just how fundamental and ambitious a revision of biological thought Bateson had intended to undertake. For *Materials*, in its structure and intention, is really a kind of medieval bestiary, a genre harking back to the very first stirrings of modern scientific thought. Bestiaries were illustrated compendia of medieval zoological knowledge in which everyday animals were depicted alongside beasts from mythology or folklore, as if the two kinds were equally real. In an age when the world outside Europe was just beginning to be explored, bestiaries were also repositories of travellers' lore and, as such, were enthralling entertainments. Shakespeare's Othello tells how Desdemona's love for him was kindled by stories of his own adventures in faraway places, in which he met an assortment of real and mythical beings:

> *of the Cannibals that each other eat,*
> *The Anthropophagi, and men whose heads*
> *Do grow beneath their shoulders.*

Parallelling the bestiaries were catalogues of human and animal monstrosities – deformities of birth. Galleries of monsters started out as freak shows, doubling as moral fables to terrify sinners into contrition. Stripped of their mystical, folkloric and moral content, these works became the first medical textbooks, as physicians began to see in monsters a way of testing the pattern of nature. If monsters were offences against nature that pointed up the normal range of variation, then monstrosities might be catalogued in a systematic way, so that the sources of this variation might be better understood.

One of the first to catalogue monstrosity in an objective fashion was Ambroise Paré (1493–1541), a contemporary of Paracelsus and surgeon to the French monarchy, whose *Monsters and Marvels* was published posthumously in 1573. Paré classified monsters according to their supposed causes, which ranged from the theological (consequences of the wrath of God) and fanciful (the result of having been cursed by a beggar) to the practical (injury or illness sustained to the mother while pregnant) and even prescient (inheritance). In 1620, the English philosopher Francis Bacon (1561–1626) recommended to his fellow examiners of nature that they make compilations of natural monstrosity as a way of understanding the boundaries of what is normal. The occasional dog or sheep born with an extra pair of spindly legs, or an extra head, serves to emphasize that farm animals should be expected to have four legs and but one head.

As historical phenomena, bestiaries and galleries of monsters can be seen as expressions of ignorance as much as early explorations of nature's plan, and they had all but disappeared by the end of the eighteenth century. The fact that *Materials* is composed along the lines of a bestiary, and has precisely the purpose laid down by Francis Bacon – that is, to attempt an understanding of the sources of variation by learning the demarcation lines between the expected and the monstrous – can be seen as Bateson's way of telling us that, despite Darwin, and indeed despite everything, our apprehension of the fundamental laws of form, generation and heredity remained as rudimentary in 1894 as it was in the days of alchemy. Bateson's moths with legs growing out of their heads are every bit as freakish and unexplained as Othello's men whose heads grow beneath their shoulders.

Bateson's emphasis on the importance of the monstrous, of gross deviation from the norm, put him at odds with the Darwinian view that change comes in very small increments. Where Darwin's mission was to explain diversity through variation whose existence was a given, Bateson sought underlying rules that could explain the disposition, and thus the existence, of the variation on which natural selection depended.

In addition, Bateson was dissatisfied with Darwin's thesis of gradual change, because the evidence of his own eyes told a different story. The variations recorded by Bateson in *Materials* tended not to be subtle matters of degree, but prominent changes in kind. There were creatures with too few digits, or too many. But there were none with fractions of digits, caught in some act of gradual atrophy or acquisition. Qualitative, disjunct variation, thought Bateson, was hardly what Darwin was talking about when he spoke of gradual evolutionary change that took geological ages to complete. Plainly, there was more to variation than Darwin suspected: it could be that the transition between species could be by way of this discontinuous variation, rather than the gradual, gentle process envisaged by Darwin.

Writing with the passion of Goethe, Bateson asserted that discontinuous variation could not be the work of selection:

> the existence of sudden and discontinuous Variation, the existence, that is to say, of new forms having from their first beginning more or less of the kind of *perfection* that we associate with normality, is a fact that disposes, once and for all, of the attempt to interpret all perfection and definiteness of form as the work of Selection. The study of Variation leads us into the presence of whole classes of phenomena that are plainly incapable of such interpretation. The existence of Discontinuity in Variation is therefore a final proof that the accepted hypothesis is inadequate.[10]

The idea of 'perfection' owes much to nature-philosophy – and nothing to Darwin. If there is any doubt about Bateson's views, he lays them to rest later on: 'it is quite certain that the distinctness and Discontinuity of many characters is in some unknown way a part of their nature, and is not directly dependent on Natural Selection at all'.[11] The phrase 'some unknown way' suggests that Bateson, like Darwin, could not explain the nature and sources of this variation. He was, like William Harvey centuries earlier, disarmingly candid about his ignorance. His critics – and there were many – could only ever take this admission as an accusation that there was something missing at the heart of Darwinism. Bateson attracted trenchant criticism from Darwin's direct intellectual descendants.

In an age in which the presentation of science was becoming increasingly quantitative and based on experiment, the self-styled inheritors of the Darwinian flame were embarrassed by Darwin's anecdotal style and sought to put his ideas on a soundly mathematical footing. If evolution were a matter of minute steps, then this minuteness had to be measurable, and so the 'biometrical' school came into being. Prominent biometricians included Galton, the statistician Karl Pearson (1857–1936), Walter Weldon (1860–1906) and Edward Poulton (1856–1943).

The biometricians reasoned that if evolution happened in very small steps, the overall effect would be one of gradual change, and that the nature of variation of a trait in a population would be 'continuous'. In other words, the discrete but slight variation between individuals could be expressed as a continuum, so that both extremes can be connected by a series of intermediates. For example, the human race as a whole shows a vast range of skin pigmentation, from almost white to deeply black. But individuals can be found with intermediate skin tones, and if you found enough individuals and lined them up in order of pigmentation, the transition from white to black would be so smooth as to appear continuous. Any idea that variation might be discontinuous, or that discontinuous variation of the monstrous kind suggested by Bateson had any part to play in evolution, was anathema to the biometricians. Bateson's almost histrionic self-importance only made matters worse: Poulton castigated Bateson for his 'narrowness, dogmatism, prejudice, and contemptuous deprecation of research about which he was regrettably ill-informed, not to mention [his] exaggerated estimation of the importance of his own work'.[12] To some, Bateson's stance against Darwin looked like betrayal: that Bateson and Weldon had once been close colleagues made the battle all the more personal.

The wounds opened by *Materials* would not close for more

than four decades, and in a sense they have not fully healed even today. Modern commentators recognize Bateson for his pioneering work in genetics, while quietly ignoring *Materials* as if it were some kind of prehistoric aberration rendered obsolete once its author had invented the science of genetics just a few years later. Bateson's arrogance and his overwrought literary style have clearly prevented his contribution from being properly recognized, but there is something else at work – something of the nature of revisionism. From the perspective of a modern Darwinian, the fact that a pioneer of genetics as important as Bateson held such passionately anti-Darwinian views is embarrassing, and so *Materials* is politely but firmly ignored.

Once this prejudice has been taken into account, and *Materials* reappraised with the perspective that only a century of distance can offer, Bateson's bestiary looks less like the irrelevant ravings of a rebel whose cause had yet to be found than a pivot around which biology turns. Between a single set of covers is a book that looks back to the very dawn of our efforts to understand the origin of form, and at the same time looks forward to genetics and the quest for the genome, whose rewards we have only now begun to enjoy.

7

Genetics

William Bateson's *Materials for the Study of Variation* was based entirely on observation. The next step was experiment, to manipulate nature directly. It was clear that inheritance had a part to play in the production and maintenance of variation; Darwin's failure lay in his inability to explain how traits were passed down from parents to offspring. Bateson set out to answer this question by undertaking a series of breeding experiments in animals and plants, looking for underlying patterns in the inheritance of traits. It is hard, now, to appreciate what a daunting task this must have been, for Bateson was faced with two significant and largely insoluble problems.

The first was theoretical – that is, he could have had no expectation about what patterns might emerge, and so no yardstick against which to judge the success of his results. In short, he was fishing in the dark. Of course, his work on *Materials*, and his general antipathy towards Darwin's idea of extremely slow, gradual change, had led him to expect that traits would be simple and discrete, and so might be detected rather easily. However, Bateson had no sure knowledge of how much more reasonable this expectation might have been, over an act of faith. The second problem was practical, and that was the choice of experimental subject. Some creatures just happen to be more suitable for breeding experiments than others, but with scant systematic work in the field there was no way for Bateson to find this out until he had chosen his subjects and worked with them for a number of years. Experiments can be ruined and theories changed by an inapposite choice of subject, and the early history of genetics was strongly influenced by the choices of experimental animals and plants made by those first geneticists.

Today, most of our knowledge of the inner workings of nature comes from experiments on just a dozen or so animal and plant species, selected from a pool of many millions for their utility in one field of study or another. By virtue of its ubiquity, the domestic chicken has been a favourite subject for studies on generation since time immemorial. More recently, an obscure species of roundworm called *Caenorhabditis elegans* became the creature of choice in experiments to test longevity, by virtue of its extremely strict, finely determined life cycle. Tabloid newspaper headlines about the elixir of youth are almost always based on the discovery of mutant worms that live for a few days longer than the three weeks or so typical for this creature. Meanwhile, a species of African frog, *Xenopus laevis*, with its big and easily visible eggs, makes a good testing ground for ideas about the very early embryo; and, as will be seen, the easy-to-breed, easy-to-keep fruit fly, *Drosophila melanogaster*, became the workhorse of the early geneticists, and would probably still top the poll in a scientific vote for Number One Experimental Animal.

These creatures have lately been joined by a handful of other animals and one or two plants, but few ask why we have chosen these particular species, rather than others, on which to build our edifice of knowledge. Practicality is an important factor,

105

followed by usage – after all, nobody is likely to start a breeding programme of blue whales to answer questions of genetics that could be answered within weeks by tests on fruit flies. But a century ago Bateson and his colleagues did not even have fruit flies to turn to, and their success owed much to their lucky choice of subject.

Bateson hedged his bets by running several different breeding programmes simultaneously, but nowhere did he find clear, unambiguous patterns in the ways that traits were transmitted, and this led to prolonged frustration. Poultry seemed a safe bet – after all, poultry breeding had been well established since Aristotle – but Bateson ran breeding experiments on several different species of animal and plant at the same time to see what turned up. Results from work on the butterfly *Pieris napi* were hard to interpret, but he appeared to make greater headway with his plants and poultry. The aims of his work were set out in a paper presented in 1899 at a conference on hybrids, but outright success eluded him. Opposition from the powerful and influential biometricians meant that he found it very hard to get funding, and had to rely to an extent on private benefactors.

Illumination came in 1900, and from an unlikely quarter – an Augustinian abbot and amateur plantsman in a relatively obscure corner of Central Europe. The cleric was Gregor Mendel (1822–84), born in Brünn (now Brno, in the Czech Republic), who received a good education in science but was, like Darwin, a college drop-out and, again like Darwin, had studied for the priesthood.[1] But there was no *Beagle* waiting to expand the mind of the young Mendel, so he took his science with him into the abbey. Mendel's voyage of exploration went no farther afield than the walls of his monastery garden, yet its consequences were just as profound, and would eventually save Darwin's faltering legacy.

Mendel's enthusiasm was plant breeding, and his favourite

subject was the garden pea, *Pisum sativum*. In 1865 he presented his results to the Brünn Society for the Study of Natural Science, the memoirs of which made their way to the shelves of more than a hundred university libraries and scientific bodies, where they remained, largely unread. After all, few would have cause to study a paper on experiments in plant hybridization published by a hobbyist in the journal of a provincial academic society. Mendel's work was not as overlooked as legend has claimed, but the fact remains that Mendel hardly went out of his way to publicize his work. It was almost as though Darwin had become a country parson; had never sailed aboard the *Beagle*; had never become a celebrated naturalist; and had published his ideas on variation as a short note entitled 'Some Observations on Natural Varieties' in his parish newsletter.

If posterity has not looked altogether kindly upon Mendel's career, he was extraordinarily lucky with the materials that were available to him for study. He wished to find out how traits were transmitted from generation to generation, and he could hardly have chosen – by happy accident – a better experimental subject than the garden pea, in which such traits are easy to measure and, as subsequent work showed, are transmitted far more straightforwardly than in many other plants or animals. Had Bateson used the garden pea, rather than butterflies, his work in the late 1890s might have progressed further.

Mendel took peas that bred true for growing tall (termed a tall 'habit'), and crossed them with plants that were equally confirmed dwarfs. According to the conventional wisdom of blending inheritance, all the progeny should have been of medium height – but they were not. They were all, without exception, tall. Even more intriguing were the results of crossing members of this first hybrid generation. Three out of four individuals in this, the second generation, were tall, and one out of four were dwarf plants, like one of the grandparents.

Mendel found similar segregation with six other traits, relating to such qualities as the texture of the peas (round versus wrinkled), their colour (green or yellow) and the way the plants flowered (along the stem, or only at the ends). In all cases he found the same pattern of inheritance: the first hybrid generation would display the trait of one or other parent – not a blend of both – and the second generation would display the trait of the first generation, alongside the grandparental trait that had become hidden in the first generation, in a 3:1 ratio.

If you had been accustomed to thinking of heredity in terms of a blending of traits, these results would have seemed incomprehensible. Mendel, however, put such worries aside, looked at the evidence with an open mind and made a great intuitive leap. For a trait – such as dwarf habit – to disappear in one generation and then reappear in the next is simply impossible according to blending, so something else must be at work. Whatever it is that carries the information about traits must remain inviolate throughout the process of generation, as a separate and distinct entity, for it could hardly 'unblend' itself and reappear after a generation of apparent absence. This would be like mixing red paint and blue paint to get purple – and having the red and blue paint separating themselves out later on according to a predictable and unvarying schedule.

But traits have their own rules about appearance. A plant would be tall, or dwarf, but not both at once. Therefore, the determinant of heredity (let's call it a 'particle') that governs stature must come in two varieties, *tall* and *dwarf*. Each parent contributes one of two possible varieties of stature particle to the progeny, which thus has two of each. If both particles are of the *tall* variety, the progeny will be tall, and would continue to breed true for that trait; if both are of the *dwarf* kind, the offspring will be dwarf. But what of those cases in which the progeny receive one each of the *tall* and *dwarf* varieties of stature particle? For some unexplained reason, the *tall* variety takes precedence over the *dwarf*, so that all the progeny would be tall (rather than short, or of intermediate stature). Some plants, therefore, would be appear tall, but would not breed true for that trait because they would contain an occult particle for shortness. So, were one to cross two of these tall plants together, the offspring would not all be tall. In fact, on average only one out of four progeny would be tall, and would breed true for tallness; one out of four would be dwarf, and would breed true for the dwarf habit; but two out of four would appear tall only because the particle for *tall* would trump the particle for *dwarf*. These plants would not breed true for tallness, as they carried the dwarf trait without manifesting it. What a nurseryman would actually see is three tall plants for every dwarf – a ratio of 3 : 1.

Mendel had no idea about the physical nature of the particles of heredity – only that they behaved as atomistic entities which were independent of the external appearance of the plants, determined by a hidden calculus of genetic etiquette in which some traits, such as tallness, were *dominant*, while others, such as the dwarf habit, were *recessive* and only appeared in the progeny in the absence of the 'tallness' variety of the particle governing stature.

Having established that traits segregated in this discrete way, Mendel set out to explore the extent to which they behaved independently of one another. He interbred plants with various combinations of different characters: he crossed tall plants that bore round peas with dwarf plants that bore wrinkled peas, and so on. Just as there is a particle governing stature, so there appears to be a particle governing pea shape – which may come in two varieties, round and wrinkled. Roundness in peas takes precedence over wrinkledness, as tallness trumps shortness. But is the stature particle a separate entity from the one that governs pea shape, or are these two hypothetical particles really aspects of the same thing? Mendel soon worked out a way to put this to the test.

If the particles for pea shape are inherited independently from those that govern stature, it is easy to predict the outcome of a cross between a true-bred tall plant with round peas and a true-bred dwarf plant with wrinkled peas. Out of every sixteen offspring, on average nine will be tall with round peas, three will be tall with wrinkled peas, three will be dwarf with round peas and just one will be a dwarf with wrinkled peas. This distinctive 9:3:3:1 pattern is exactly what Mendel found, good evidence that the particles for stature and pea shape were passed on independently. Mendel found this result repeatedly for combinations selected from seven different traits, including flower position and seed colour. Whatever it was that determined heredity – essences, particles, atoms, gemmules – they behaved discretely and independently, just like different-coloured balls in a bag.

Mendel was lucky not just in his choice of experimental subject, but also in his choice of traits. As later events showed, most traits in most animals and plants do not segregate independently. Had some of the traits – the colours and textures of peas, say – not sorted independently, the results would not have been as clear cut, and Mendel would have found himself in the same position as Bateson, unable to find a clear explanation for his results. But, as any bright-eyed holder of a lottery ticket will tell you, the chances of winning the big prize may be minuscule, but *somebody* has to win it.

Other researchers referred to Mendel's published work over the years, but nobody grasped its significance, or that it showed how to address the nature of variation. Bateson grew up in ignorance of it, and in his *Materials*, almost thirty years after Mendel had published his paper, he bemoaned the lack of experimental work to elucidate the nature of variation. When

Darwin died in 1882 it was an occasion for national mourning, and he was buried in Westminster Abbey, traditionally the last resting place of the greatest of Britons. Mendel died two years later, and although much mourned in Brünn (the composer Janáček played the organ at the funeral), his passing was hardly the talk of Prague or Vienna.

When Bateson was investigating the nature of inheritance in butterflies and chickens, a Dutch botanist, Hugo De Vries (1848–1935), was studying a plant, the evening primrose (*Oenothera lamarckiana*), which had the curious ability to produce sports — monsters — which subsequently bred true. Here is another instance of theory being dictated by choice of subject: the fact that sports of *Oenothera* breed true is now known to result from a quirk of genetics peculiar to that plant, but De Vries used his discovery to develop a general theory in which new species arose from such sudden, unpredictable events, which he called 'mutations'. Nature, then, could create new species all at once, without the need for gradual change. We now know that random mutation is vital for generating the variation on which natural selection can act, but what we mean today by the term 'mutation' is rather different from what De Vries had in mind, which at the time seemed quite at odds with the Darwinian idea of gradual, continuous change.

To explain his results, De Vries came up with a mechanism of heredity he called 'intracellular pangenesis', which was similar to Darwin's idea in that the number of 'pangens' (equivalent to Darwin's 'gemmules') contributing to a trait was variable. Investigating this, he crossed plants with contrasting traits (hairy versus smooth stems, for example), interbred the progeny, and achieved ratios between the contrasting traits, in the second cross, of three plants with one trait (such as hairy stems) for every one plant with the contrasting trait. In a series of three papers published in 1900, De Vries set out to explain this curious

111

3:1 ratio. Each pair of contrasting traits (hairy or smooth stems, for example) was associated with a single pair of pangens, and not a variable number as he had thought they might be. During reproduction, each parent passes on just one of each pair of pangens for any given trait. No sooner had De Vries published this work than he received papers from other botanists claiming that they too had found the 3:1 pattern of inheritance in their own experiments. In fact, everyone had already been beaten to it by Mendel, who had published his elegant work on garden peas a generation earlier, only no one had appreciated its significance.

According to biological legend, the news about Mendel reached Bateson in 1900 while he was on a train from Cambridge bound for London, where he was to give a lecture to the Royal Horticultural Society.[2] He had reported progress in his breeding experiments the year before, and had planned to give another report along the same lines. As ever, progress had been slow and the results ambiguous, so Bateson was not looking forward to the occasion with much eagerness. Having been sent one of De Vries's papers in which Mendel's work was mentioned, Bateson sought out a copy of Mendel's paper and took it with him to while away the train journey. As he read it, Bateson realized that Mendel had found the mechanism of inheritance that he himself had been fumbling towards with great effort and only modest success. By the time the train reached London, he had decided to abandon his planned talk in favour of a discussion on Mendel's long-neglected research. That lecture was the first notice of Mendel's work in Britain, and because Bateson was the herald, Mendelism – a scheme of discrete and distinct particles of heredity – was raised in explicit opposition to the biometrical school of continuous variation. Here, Bateson thought, was the work that proved him right and the biometricians wrong. With characteristic fervour, he translated Mendel's work into English and coined the term 'genetics' for the study of the particles of heredity, the particles themselves becoming known as 'genes'.

At that time the physical nature of genes was unfathomable. The first geneticists regarded genes less as real things than as manifestations of some as yet undefined underlying principle that governed the transmission of traits and the maintenance of variation. Despite their other differences, the biometricians and the emerging school known as the Mendelians all thought of genes in the way that economists think of money – a convention for the comparison of value, but with only an abstract existence of its own. Both groups adhered to the need for cautious, sound experiment rather than speculative theorizing, and were uncomfortable in suggesting that genes had a physical existence. The biometricians were probably still aware of Darwin's rather self-conscious venture into gemmules, and possibly had not yet forgotten how Buffon had made himself a laughing stock with his elaborate theory of atomistic inheritance. This purist stance led to a lot of confusion between traits (the visible expression of the action of genes) and the genes themselves. In retrospect, the conclusion that genes are real things is unavoidable, but the early geneticists went to great lengths to avoid drawing it.

One visitor to De Vries's laboratory and garden in Hilversum in Holland in 1900 was Thomas Hunt Morgan (1866–1945), an American embryologist who was just about to take up a position at Columbia University in New York. From Hilversum the tourist went south, to the famous marine laboratory at Naples, where Mendel's results must have been the latest sensation. After his visit to De Vries, Morgan wrote that biology was on the verge of producing a new theory of evolution in which species were created outright. Morgan was disillusioned with Darwinism: De Vries's 'mutation' was the brave new theory.

Morgan was born in Lexington, Kentucky, where his fame is still eclipsed by that of his swashbuckling uncle, John Hunt Morgan, the 'Thunderbolt of the Confederacy'.[3] Thomas got his first degree, and his master's degree, in Kentucky, before getting out of his home state, more or less for good. He completed his doctorate at Johns Hopkins University – where he, like Bateson, had come under the influence of William Keith Brooks – before moving to Bryn Mawr, and then Columbia. Like Bateson, Morgan was interested in marine creatures such as acorn worms, the study of whose anatomy and embryology it was hoped would furnish clues to the heritage of the vertebrates. He wrote extensively on the development of frogs, and investigated the remarkable properties of regeneration in simple animals – the same phenomenon that had captivated Trembley in the 1840s, and had turned the young Albrecht Haller away from spermism. Like Bateson, Morgan had been a firm Darwinist before his disaffection. His road back to Darwin – guided by his students – was hesitant and grudging. Again, like Bateson, he was no fan of idle theorizing, and always had several experiments in progress, testing notions founded in embryology, physiology and generation. All his experiments, he said, were at best foolish, and most came to nothing. Such is the way of science, then as it is now.

One such experiment started in 1908, when Morgan set a student, Fernandus Payne, to test the Lamarckian idea that species could change through the acquisition of characters – that an animal's response to its environment could be inherited by its offspring. Now, it may be true that the children of people who acquire great wealth will be born wealthy – but it is less clear whether, say, giraffes who spend their lives straining their necks to reach higher branches will endow their offspring with yet longer necks. Yet such was the claim of Lamarck, and it was still taken seriously at the turn of the twentieth century, when Darwinism was at its nadir of popularity for want of a mechanism to explain the origin of variation.

Payne was interested in how and why animals living in total darkness, in subterranean caves, for example, are often blind, or even eyeless. The Darwinian explanation would be that in total darkness animals with fully developed eyes would be at a disadvantage relative to blind or eyeless ones, as they would have devoted resources to the development of what were, in effect, useless organs. Given the existence of natural variation in the sightedness of individuals in the population, natural selection would favour animals with lesser rather than greater capacity for vision, and over many generations the proportion of blind individuals would increase. The Lamarckian explanation, which would have seemed equally plausible in 1908, was that sighted animals raised in darkness would experience degeneration of their eyes in their own lifetimes, and would pass this atrophy on to their offspring.

Whereas Bateson had his butterflies and Mendel his peas, Payne's experimental animal of choice was to prove the most enduring and influential of any organism in genetics. Enter the fruit fly, *Drosophila melanogaster*, first used a few years before in experiments on inbreeding. These tiny flies turned out to be the perfect experimental animals. They are small, breed prolifically, are relatively easy to handle (they can be anaesthetized with a puff of ether and studied with a hand lens), are easy to feed,[4] and large numbers can be kept in a small space, such as a milk bottle. Payne raised sixty-nine generations of flies in complete darkness. On emerging into the light, the sixty-ninth generation flew straight towards a lighted window, clearly just as sighted as the ancestor that had first gone into the dark. The experiment proved nothing, but such loose ends were all in a day's work for Morgan and his students in the cramped and smelly laboratory at Columbia that came to be known as the Fly Room.

Not to be deterred, Payne and Morgan embarked on another experiment, this one inspired by De Vries's suggestion that mutations might be artificially induced by chemicals, or by the recently discovered phenomenon of radiation. The researchers spent two years exposing flies to X-rays, wide ranges of temperatures and an arsenal of chemicals, but the flies proved as resilient to these insults as they had to darkness. 'There's two years work wasted,'[5] complained Morgan in 1910, as he showed off his lab to a visiting former colleague from Bryn Mawr.

But one day early in 1910, a mutant fly turned up at the laboratory. This fly had white eyes, not the usual red ones. The provenance of this fly has long been debated. Morgan believed that it appeared in early January (coincident with the birth of his third child) and could have been a product of the mutation experiments. However, a colleague from the Cold Spring Harbor Laboratory claimed that he had donated it to Morgan – a claim that Morgan could not refute with certainty. And of course, it could have simply flown in through a window. Or there could have been several such flies. Such a muddle was probably to be expected given the haphazard organization of the Fly Room, where several experiments were going on at once in a small space.

It is possible that mutant flies really did appear in Morgan's laboratory as a result of the mutation experiments before 1910, but had been missed, because Morgan and his colleagues were not sure what they were looking for. The very idea of mutation was quite new, fruit flies are extremely small, and variations in, say, the number of bristles on a fly's leg might not have been scored as mutations, even had they been noticed. However, there certainly was a white-eyed fly in May 1910, when Morgan wrote to Bateson to tell him about it. After much disappointment, he was finally on to something. Once the first mutation had been recognized, others followed in quick succession. Flies with pink or vermilion eyes joined the white- and red-eyed varieties. Flies were found with unusual colour patterns and deformities to their wings or limbs. By the end of 1912, forty different mutants had been catalogued.

At first, the flies were almost as well behaved as Mendel's peas. The lone white-eyed fly was perforce mated to a red-eyed fly. The offspring all had red eyes, but when the flies in that generation were interbred the result was 3,470 red-eyed and 782 white-eyed flies, a ratio of almost 4.5 : 1. A deficiency of white-eyed offspring meant that this was not quite the Mendelian 3 : 1 ratio, but the result was clear. The white-eye trait was recessive, and in no case had there been blending. But why were there fewer white-eyed flies than expected? It turned out that the ratio was strongly influenced by gender – white-eyed flies were far more likely to be male than female. Morgan concluded that the gene for eye colour was somehow linked with whatever it was that determined the sex of the fly. Unlike Mendel's traits in peas, some (but not all) of the genes in flies interfered with one another, skewing the proportions of offspring with particular traits away from the Mendelian expectation that traits would be passed on independently of one another. Bateson, working on poultry, had come to the same conclusion – some genes tend to be passed on independently, whereas others are associated, or linked.

The white-eyed trait was not the only one that was linked to sex. Morgan and his colleagues got similar results for flies with vermilion eyes, and with a mutation that had abnormally small wings. As work went on, it became clear that all the genes in an organism could be allocated to a smaller number of *linkage groups* in the same way that all the individual footballers in a league can be allocated to a smaller number of teams. Traits whose genes shared a linkage group would not be inherited independently, and this would be manifested in a greater or lesser deviation from Mendelian ratios. In contrast, traits whose cognate genes – ones directly associated with the trait – were members of separate linkage groups behaved independently, as had Mendel's traits in peas.

This phenomenon illustrates Mendel's good luck extremely well. As it happens, each of the seven traits of *Pisum sativum* studied by Mendel can be allocated to a distinct linkage group. By sheer good luck, Mendel had examined just one trait from each of seven linkage groups – in no case did he report traits associated with genes in the same group. Had he done so, then the proportions of offspring of different kinds would not have been as expected, and would not have been clear – perhaps no clearer than Bateson's results with birds and butterflies. Mendel, however, hit the jackpot at once: there was no mistaking that the ratios he saw represented a genuine pattern in nature, and other geneticists could then use Mendel's ratios as a yardstick against which to judge their own results[6] Without Mendel's results, it would have been impossible to understand the distribution of traits in terms of linkage groups. As Morgan and his associates performed more breeding experiments on mutant fruit flies, several linkage groups began to emerge from their data. For example, the traits for white and vermilion eyes, miniature wings and the determination of sex all appeared to belong to one linkage group. The trait for pink eyes, however, segregated independently of these three traits, and belonged to another linkage group. By 1914, Morgan's team had concluded that there were four separate linkage groups in *Drosophila*.

Work on the association of genes in linkage groups revealed a degree of order within each group. It seemed that the gene for each trait had a definite 'location' in the group that could be 'mapped' relative to that of any other in a consistent way. It turned out that the disposition of genes in any linkage group could best be understood in linear terms, as if they were so many birds on a wire, each occupying its own station relative to that of the others. The notional 'distance' of each gene from any other could be defined by the degree to which they were linked – that is, the degree of deviation from expected Mendelian ratios, as judged from breeding experiments involving the two traits associated with the genes. For traits associated with small deviations from the Mendelian ratio, the genes associated with those traits would lie far apart on the linear map of the linkage group. For traits associated with larger deviations, in contrast, the cognate genes would lie closer together. As they gathered more and more data from breeding experiments with an ever larger array of mutant flies, Morgan and his team were able to make what were, in effect, the first maps of a genome. Morgan set out his ideas on linkage and genetic mapping in 1915.

It cannot be stressed too strongly that, to Morgan and his colleagues, none of this represented reality any more than a map of California is actually made of the sunshine and palm trees of Pacific Grove, where Morgan liked to spend his summers. For this reason, a gene which has not been physically isolated but is suspected – from breeding experiments, or analysis of pedigrees – to lie in a certain position on a linkage map was called a *locus* in order to emphasize geography over substance. Morgan went to great lengths to emphasize that this elaborate cartographical arrangement was nothing more than an abstraction. The distinction between gene and locus is still made today, to distinguish between the location of a gene of interest and the gene itself.

Alongside the early geneticists trying to deduce the rules of inheritance from breeding experiments were the cytologists – cell biologists – who took a more straightforward approach: they examined living tissue directly, to find out what could be learned of the physical basis of inheritance. Intellectually, the cell biologists were the direct lineal descendants of embryologists, such as Von Baer (who described the human ovum in 1828) and, through them, the more practically minded of the nature-philosophers who sought to understand the pattern of nature by copious observation of the real world, rather than an idealization of it. Throughout the nineteenth century, microscopy had advanced to such a degree that it was now possible to observe structures inside the nuclei of cells. Chief of these were bodies called chromosomes (from the Greek for 'coloured bodies', an allusion to the fact that they were readily stained with various preparations to improve their visibility through the microscope).

Chromosomes are fascinating structures whose behaviour must at first have seemed as strange and arbitrary as that of the Cheshire Cat in *Alice's Adventures in Wonderland*. All cells have chromosomes, except that sometimes there do not seem to be any there at all, and at other times there are twice as many as expected. Chromosomes in cells which are about to divide appear to engage in an elaborate and immemorial choreography that almost never varies, irrespective of the organism in which it takes place. Decades of patient observation by many researchers eventually revealed a consistent pattern to chromosome behaviour.

Nearly all nuclei from the cells of a given species have a chromosome number which is characteristic of that species. These chromosomes are generally arranged in pairs. Human beings have forty-six chromosomes in twenty-three pairs. In twenty-two of these pairs the members of each pair look identical, although each pair might have a distinct appearance, but in the twenty-third pair, the so-called 'sex' chromosomes, the partners constitute the tiny Y chromosome and the much larger X chromosome. Chromosome number is no guide to the position of an organism in any subjective *scala naturae*: although the humble fruit fly has only four pairs of chromosomes, goldfish have fifty-two pairs – more than twice as many as in humans.

A cell reproduces by dividing into two daughter cells, and it is during the process of division that chromosomes come to the fore. Before division, the chromosomes are long, thin structures stretched so fine that they are all but invisible. During cell division each chromosome (in each pair) makes a precise copy of itself; after division, the chromosomes shorten and become thick enough to be visible using an ordinary microscope. Also during cell division the nucleus splits into two, arranging matters so that each chromosome becomes separated from its copy, ensuring that each new nucleus has precisely the same number – and same kinds – of chromosomes as had the parent nucleus. This is the routine kind of cell division that occurs in most tissues, most of the time, and it is called mitosis.

A rather different kind of cell division, called meiosis, happens in those tissues of the sex organs that produce sperm or ova. Cells whose descendants will become sperm or eggs undergo a sequence of divisions in which the chromosomes first copy themselves, as in mitosis, except that there is an extra stage, called recombination, in which the duplicated chromosomes in each chromosome pair line up to form four-handed structures, called chiasmata, and randomly exchange pieces of themselves with one another. Recombination ensures that the resulting sex cells will be as genetically variable as possible.

Recombination is notoriously hard to describe in words, so something more graphic may help, and for convenience I shall advance the argument a stage further, presenting the idea that the age-old dance of the chromosomes explains the otherwise arbitrary rules and behaviour of genes, as inferred from breeding experiments. Imagine a chromosome along which loci (as distinct from genes) are arranged in linear order. Here, a letter stands for a locus:

RECOMBINATION

Each chromosome is a member of a pair. The other member of the pair has the same loci, in the same order – but in different varieties, or *alleles*:

recombination

Alleles correspond to different varieties of the same trait. The 'T' in one chromosome and the 't' in the other could stand for different alleles of the stature trait in Mendel's peas, with 'T' representing the dominant *tall* trait and '**t**' standing for the recessive, *dwarf* trait.

In the first division of meiosis, the chromosomes copy themselves:

RECOMBINATION
RECOMBINATION

recombination
recombination

and then line up together to form a four-handed chiasma:

RECOMBINATION
RECOMBINATION
recombination
recombination

This is when recombination takes place. Portions of the four chromosomes 'cross over', or recombine. In this example, the chromosome at the top of the list exchanges material with the third on the list:

RECOMBination
RECOMBINATION
recombINATION
recombination

The crossing points are random, and more than one crossing-over may occur in any given chiasma, involving any or all the chromosomes. After this process, which can actually be observed under the microscope, the cells divide and divide once again to become four cells, each with only one chromosome in each pair – that is, half the normal complement of chromosomes found in most cells in the body:

RECOMBination

RECOMBINATION

recombINATION

recombination

In females, only one of these four cells matures, to become an ovum. In males, all four have the potential to become mature sperm.

To backtrack slightly to before the birth of genetics, cytologists in the last quarter of the nineteenth century suggested that the consistent behaviour of chromosomes during processes such as recombination had parallels with the laws of inheritance. As long ago as the 1880s, the German biologist Theodor Boveri (1862–1915), working on the horse roundworm *Ascaris megalocephala*, noticed that chromosome numbers halved in the formation of sex cells. When he read De Vries's account of Mendel's laws, Boveri immediately saw how the basic behaviour of chromosomes was too much like that of De Vries's paired particles of inheritance, his 'pangens', to be a coincidence. The chromosomes came in pairs, reasoned Boveri, so you could think of pairs of De Vries's pangens, one of each pair residing on one chromosome, and the other – representing the contrasting trait – residing on the other chromosome. The chromosomes of one generation were apportioned equally between both parents, and passed on to the offspring, in a manner consistent with De Vries's pangenesis and Mendelian laws.

Boveri's work was confirmed and extended by the American biologist Walter Sutton (1877–1916), working at Columbia on the cells of the grasshopper *Brachystola magna*. The paired chromosomes in this species have distinctive shapes, so once you have spotted one of each pair on a microscope slide, it is easy to identify the other – almost as easy as counting different species of creature on to Noah's Ark, two by two. Sutton discovered that the sex cells broke this rule: each had only one member of each pair of chromosomes, and the full complement was restored only after fertilization, when sperm and egg united and fused.

Summing up this work in 1902, Sutton concluded that the behaviour of chromosomes could constitute the physical basis of heredity, as played by Mendelian rules. The next year, he made an explicit link with Mendel in a paper entitled 'The chromosomes in heredity'.[7] Sutton's work appeared before Morgan arrived at Columbia in 1904, but Morgan was impressed by the advances in cytology that were being made at Columbia, and their significance was not lost on him. But still he hesitated – if cytologists made bold claims about the connection between chromosomes and Mendelian laws, that was their business. The geneticists might have only been next door, or down the hall, but they were liable to be more cautious than the cytologists when proposing elaborate theories of the physical reality of genes. The ignominy of Buffon – and Darwin – still rankled.

It was recombination that finally changed Morgan's mind. He and his colleagues had explained how deviations from the expected Mendelian ratios, observed in thousands of breeding experiments, could be explained by pooling loci into linkage groups, and that linkage groups could be seen as maps in which loci were arranged in a fixed, linear order, like stations on a railway line. This theoretical arrangement explained the results extremely well, but as Morgan and his colleagues created more and more elaborate linkage maps, it became less and less easy to dismiss genes as abstractions. Were linkage maps a reflection of reality, or simply delusions built on delusions – like Percival Lowell's ever more detailed maps of the canals of Mars, later shown to be illusory – or, worse, soap-bubbles of bold yet groundless assumptions, no better than Buffon's atomistic theories of inheritance? The behaviour of chromosomes provided a ready explanation of the behaviour of genes, but only if you admitted that they were real – and recombination provides the most telling test of all. It turned out that the theoretical process of mapping loci in linkage groups was actually a consequence of the process of recombination during meiosis, in which the members of a chromosome pair, each of which has just replicated itself, form a four-stranded structure for the purpose of swapping genetic material.

Recombination is a physical process, and takes up real-world, physical space. Crossing-over is not an abstract invention, but an actual phenomenon in which chromosomes are cut and exchange parts of themselves with one another. Looking back at my representations of recombination on pages 122–3, you'll see that two of the chromosomes split between the 'b' and the 'i' of 'recombination'. However, there is no reason why the split should occur at *precisely* this point, and in fact it can happen at any random point along the length of a chromosome.[8] Because of this randomness, there is more chance that the split point will be between loci that are widely separated on a chromosome than between loci that are closer together. By way of analogy, there is more chance of winning at roulette if you simply bet on red or black, rather than piling your chips on just one number. In the same way, you are twelve times as likely to stick a pin in the gap between *any* two adjacent letters in the word RECOMBINATION than between any *specified* pair, such as the 'M' and the 'B' in the middle.

Because crossing-over points are less likely to occur between any specified pair of closely linked loci than between loci spaced farther apart, these closely linked loci will tend to be inherited together more than might be expected, were they widely spaced or even completely unlinked. This means that traits associated with loci that lie close together in a linkage group will have less chance of being split up during a crossing-over event, and will therefore not be inherited according to Mendelian expectation — which assumes that loci are unlinked, as physically independent of one another as rabbits in a sack. The result, in a breeding experiment, is the deviation from Mendelian ratios that Morgan had observed, and — this is the clincher — the deviation gets progressively greater as linkage gets closer. From this, it is easy to see that linkage groups are merely the theoretical expression of chromosomes, and furthermore of chromosomes in which loci are arranged in precisely the linear order that Morgan and his colleagues had predicted on the assumption that loci were entirely abstract. Finally, Morgan had to concede that genes were real, physical entities. Buffon had been at least partly right. The stuff of inheritance was real, and the quest to find the entity that entices form from the formless was about to enter a new era.

Bateson died in 1926, and went to his grave deeply sceptical about the physical reality of genes. In 1933, Morgan received the Nobel Prize for Physiology or Medicine in recognition of his contribution to the chromosome theory of inheritance – an idea that he himself accepted only rather reluctantly, and only then after the strenuous efforts of his students to persuade him. These students – Alfred H. Sturtevant (1891–1970), Calvin B. Bridges (1889–1938) and Hermann Joseph Muller (1890–1967) – became influential geneticists in their turn, and Muller won a Nobel prize in his own right in 1946 for his work on artificially induced mutations. Another, Theodosius Dobzhansky (1900–1975), fused laboratory-based genetics with field natural history – learned in his native Russia – to show how the new genetics could be used to explain the kind of natural variation that intrigued Darwin. In his book *Genetics and the Origin of Species* (1937), Dobzhansky finally used the new genetics to bring Darwinian natural selection in from the cold. His work forms the basis of what is known as the 'modern synthesis' – the understanding of Darwinism that we have today.

Morgan left Columbia in 1928, moving to California full time to head the biology programme at the California Institute of Technology in Pasadena. Sturtevant, Bridges and Dobzhansky followed Morgan, pursuing genetics, while Morgan returned to the studies of regeneration that had captivated him before his fateful meeting with Mendelism. Under Sturtevant's direction, the laboratory at Caltech continued to produce work of the first rank in both experimental embryology and genetics, as it does to this day. Morgan had a right to be cautious about the physical reality of genes. As a confirmed experimentalist, he could still ask one last, unanswered question. Even were one to accept that genes had a physical existence, what were they made of?

What *was* the substance of heredity, the stuff of life?

PART TWO

8

'It has not escaped our notice . . .'

Now that chromosomes had been identified as the seat of inheritance, the next task was to find what it was that chromosomes were made of, and how this substance could act as the repository for genetic information. It quickly became apparent that any substance charged with carrying the information of heredity would have to fulfil exacting specifications. It would have to encode a large amount of information in its structure, and be accommodated in a small space – the cell nucleus.

The story of the identity of this special substance began in 1869, in a laboratory in a castle in the German city of Tübingen, where one Johann Friedrich Miescher (1844–95) discovered nuclein, the substance we now know as deoxyribonucleic acid. Miescher came from a family of distinguished scientists and physicians, but an early career in medicine had been thwarted by an attack of typhus that had left him partially deaf. Turning to chemistry instead, he set out to investigate what cells were made of. His unpromising materials were white blood cells found in pus washed from the bandages of casualties from the Crimean War.

At the time, cells were thought to be made of protein, the

ubiquitous substance of life. The word 'protein' was coined in the 1830s, and proteins have since been found in all living tissues, in an impressive variety of forms. Our hair and nails are made of a protein called keratin; the collagens that comprise the connective tissue that stick us all together are also proteins, and the substances that create blood clots are derived from protein; haemoglobin, the substance that carries oxygen in the blood, is based on protein, and so are many hormones that we cannot live without, such as insulin.

Miescher, however, found in white blood cells a substance entirely different from proteins. This mysterious and unsuspected substance was resistant to the enzymes known to digest proteins. It contained an unusually large amount of the element phosphorus, which is not usually found in proteins. Most significantly for Miescher, this substance was especially abundant in the nuclei of the cells and not very common outside them. Proteins, in contrast, were found in all living tissues – including, it has to be said, nuclei. However, because of its peculiar location, Miescher called this new substance 'nuclein', and in 1874 he speculated that it might be involved in heredity in some way. He did not follow up his own suggestion, however, and nuclein – re-named nucleic acid in 1889 and now largely synonymous with the genetic material deoxyribose nucleic acid, or DNA – languished on the sidelines, a chemical curiosity.

Proteins, however, remained of absorbing interest to scientists. The versatility of proteins comes from their molecular structure. All proteins are polymers – modular in construction, made from smaller molecular units called amino acids which are arranged in linear sequences like carriages in a train. The nature of the protein is determined by the order in which the amino acids are assembled. Like the twelve notes in the musical scale, which can be arranged to create every tune ever written, whether composed by Mozart or Motörhead, a limited selection of around twenty amino-acid units can be arranged in countless different ways to create a potentially unlimited range of molecules, each with a unique set of properties. If proteins form the basis of living tissue, from cartilage and keratin to haemoglobin and muscle, there seemed no reason why proteins too should not be the carriers of genetic information. Indeed, the modular, 'digital' construction of proteins looked ideal for encoding information in the succinct way that genetics demands – indeed, proteins had the potential to encode far more information, with greater economy of space, than did nuclein.

Like proteins, nucleic acids such as DNA are polymers, made up of a set of smaller elements in the same way that words are made of letters. Whereas proteins are made from amino acids, DNA is made from units called nucleotides. Each nucleotide in DNA consists of a substance called deoxyribose phosphate, attached to another molecule known as a base. The deoxyribose phosphate molecules link themselves together in a chain, the bases hanging from each unit like charms on a bracelet. The presence of phosphate in each and every nucleotide explains why Miescher found so much phosphorus in nuclein, but it is the bases that make each nucleotide distinctive. However, unlike the twenty amino acids commonly found in proteins, the selection of nucleotides available to make DNA is much smaller – just four different bases can be found in DNA. The bases are called cytosine, thymine, adenine and guanine, abbreviated to C, T, A and G. But matters are simpler still: adenine and guanine are similar to each other, belonging to a class of molecules called purines. Cytosine and thymine are also very similar to each other, and are both members of a class of molecules called pyrimidines.

When compared with the amino-acid alphabet of proteins, the four-letter alphabet of DNA seems poor indeed. For example, 1,280 million possible seven-letter 'words' can be made with an alphabet of twenty letters – the equivalent of roughly one word for every person in China. Reduce the choice of letters to four, however, and the number of possible seven-letter words shrinks dramatically to 16,384, the population of a small town. A protein molecule made from a given number of amino acids thus has vastly greater scope for containing information than a DNA molecule of the same length. The cramped conditions of the cell nucleus, together with the enormous amount of genetic information responsible for prescribing even the simplest cell, means that information must be encoded with the maximum possible economy. And that was the reasonable argument that prejudiced scientists against DNA as the carrier of genetic information. Until the 1950s, many thought that proteins, with their great diversity, would carry genetic information, and that the rather less versatile DNA would serve some kind of structural purpose – perhaps as inert scaffolding on which the proteins would be draped. Of course, the question of the identity of the genetic material would be resolved not by arguments such as this, but by experiment. The results came in 1944, when the American biologist Oswald Avery (1877–1955) and his colleagues established beyond doubt that, for all its inadequacies, the genetic material was indeed DNA.

In 1913, Avery joined the Rockefeller Institute Hospital in New York to study pneumococcus bacteria, which cause a variety of serious infections including pneumonia (Avery's speciality) and meningitis. Like all organisms, pneumococci contain proteins and nucleic acids, as well as other substances such as sugars. Avery and his colleagues discovered that sugar is something on which the bacteria depend for their existence, for without a sugar coating bacterial cells are incapable of causing an infection. The researchers were able to breed pneumococci in the laboratory, just as Morgan's team had bred fruit flies. And, as with fruit flies and their various eye colours, it was possible to isolate various mutant or defective strains of the bacterium. One of the mutant strains lacked a sugar coat, and although alive was incapable of causing infection. The researchers suspected that this mutation was caused by an alteration in a gene – a gene responsible for making sugar, perhaps – but the physical nature of that gene remained obscure.

However, there was a way of using pneumococci to shed light on the problem of the physical nature of genes. Bacteria are essentially scavengers, and will readily assimilate the remains of dead neighbours in a laboratory Petri dish. Live bacteria without their sugary coats were not infectious, and neither were heat-killed bacteria that retained their envelopes even in death. A mixture of the two kinds of bacteria when cultured together should have remained as uninfectious as either of its two constituent parts – unsugared and alive, and sugared but dead. But this was not so. The researchers found that mice inoculated with this mixed culture suffered the symptoms of live, pneumococcal infection. The simple explanation for this curious result is that the live, sugar-free bacteria stripped the coats from their dead neighbours and so became infectious. The more subtle interpretation is that the live, sugar-free bacteria had obtained genetic material from the dead, sugared bacteria, and used this information to make sugar coatings for themselves. Avery and his team suspected that this genetic material might be DNA.

At this point, Avery and his colleagues designed an ingenious experiment that exploited Miescher's finding that DNA contains large amounts of phosphorus, whereas proteins contain relatively little of this element. Like most chemical elements, phosphorus can be found in a number of varieties, or isotopes, some of which are radioactive to a greater or lesser extent. Bacteria feed on radioactive phosphorus (in this case a mildly radioactive isotope designated phosphorus–32) as eagerly as any other kind, so when the researchers added phosphorus–32 to a culture of live, infectious bacteria, the bacteria soon built it into their DNA. Radioactive substances can be tracked by exposing photographic film in their presence – this, after all, is the basis of radiography – so it was easy to spot the bacteria that had fed on the radioactive phosphorus. These radioactively labelled bacteria were then killed, and, still radioactive, were mixed in with a culture of the mutant strain that was still alive, but which lacked a sugar coat. Using radiography, the scientists were able to follow the DNA as it was scavenged from the remains of the dead bacteria by the live, sugar-free ones. The live bacteria were then able to create their own sugar coats and become infectious. By showing that DNA was the substance that was transferred between one bacterium and another, Avery's team proved that DNA was the material in which genetic information resided.

Establishing DNA as the material repository for genetic information was a great achievement. When Avery and his colleagues made their discovery, nothing was known about the precise structure of the DNA molecule – how the bases were arranged on a sugar-phosphate backbone. Working out this structure was the next step.

The leap from substance to structure was made possible by the discovery of X-rays by Wilhelm Röntgen in 1895, and the insights of a remarkable father-and-son team: William Bragg (1862–1942) and his son Lawrence (1890–1971). In 1912, the German scientist Max von Laue (1879–1960) and his colleagues discovered that when X-rays were shone through crystals of various kinds, distinctive patterns of light and shade were produced by the X-rays being diffracted as they passed through the regular lattices of atoms that made up the crystals. Von Laue had predicted this effect but was unable to explain precisely why lattices of certain forms produced the diffraction patterns they did, whereas apparently closely similar lattices produced very different patterns. William Bragg, who was then Professor of Physics at the University of Leeds, discussed the results with his son, a twenty-two-year-old student at Cambridge. The outcome was that Lawrence Bragg developed and refined Von Laue's work, establishing a clear and predictable correspondence between the disposition of atoms in a crystal and the diffraction pattern made when X-rays were shone through it. Thus the science of X-ray crystallography was born with Lawrence Bragg as its leading exponent, and the technique was used to elucidate the structures of many minerals and metals, before being applied to the crystalline forms of large biological molecules such as vitamin B-12, haemoglobin, myoglobin,[1] and the regular protein coats of some simple viruses. Although interpreting X-ray diffraction patterns remains notoriously difficult, the method is used to this day to determine the structures of large biological molecules.

DNA was just such a molecule, and solving its structure by X-ray diffraction was the aim of the biophysicist Maurice Wilkins (b. 1916), working at King's College in London. In 1951 he was joined by Rosalind Elise Franklin (1920–58), already a respected talent in the field of X-ray crystallography. In the same year, a young American zoologist called James Watson (b. 1928) moved to the Cavendish Laboratory in Cambridge, where members of Lawrence Bragg's research group were making great advances in solving the structures of various proteins. Watson, a precocious talent who had enrolled at the University of Chicago at the age of fifteen, had become enthused by Avery's experiments. Visiting the zoological station at Naples – the destination of another promising American talent half a century earlier – he attended a lecture in which Wilkins discussed the structure of DNA and presented the latest

X-ray diffraction photographs. Watson saw that determining the structure of DNA would be essential to understanding its function as a genetic material. Once at Cambridge, Watson met Francis Crick (b. 1916), a physicist turned biologist who was working in that part of Bragg's team engaged on solving the structure of haemoglobin.

Crick had worked on developing magnetic mines for the Navy during the Second World War, but a keen nose for important problems had led him to biology, and to the technical challenges posed by unravelling the structures of biological molecules such as proteins and nucleic acids. Crick had been intrigued by the work of the American scientist Linus Pauling (1901–94), who had discovered that many proteins have the form of a helix – like molecular corkscrews. When Watson arrived in Cambridge full of enthusiasm for DNA, he and Crick began to make models of DNA based on Pauling's helices, aided by improved X-ray pictures now emerging from Franklin – whose unpublished work was used without her knowledge or permission. Watson and Crick's search for the perfect model of DNA was driven by likely competition from Pauling, and the events of the period are described in Watson's own lively and idiosyncratic account, *The Double Helix*.[2]

At first sight, DNA is so simple, at least when compared with proteins, that there can be only a very small number of ways in which bases could be arranged on a deoxyribose phosphate strand. However, several different structures were proposed in the early 1950s. In all of them, DNA was seen as consisting of one or more long strands of deoxyribose phosphate from which bases were suspended, but the differences lay in the relationship between the DNA strands.

In one proposal, from Pauling, DNA was held to consist of three separate strands, twisted round one another to form a triple helix – a chemical version of the Three Graces. The three deoxyribose phosphate chains would form the core of the DNA fibre, with the bases projecting outwards like the bristles of a bottle brush. Given the iconic status that the DNA double helix now enjoys, a proposal for a triple helix may now seem strange, except that at least one protein, collagen, is known to have a three-stranded form. Another proposed structure also had three strands, but with the backbones on the outside and the bases pointing inwards. Still other models had just two strands. Using models as well as Franklin's data, Watson and Crick devised a DNA structure with two deoxyribose phosphate strands winding around a vertical axis, with the bases projecting inwards from each strand, towards the axis, meeting one another in the middle.

The Watson–Crick structure explained a curious feature of the chemistry of DNA. In 1952 it had been found that the bases in DNA are always present in a certain ratio. Were you to break up any strand of DNA into its constituents and count the bases, you would find that the number of cytosines is always the same as the number of guanines, and the number of adenines always equals the number of thymines. Watson and Crick's structure accounted for this curious property in the following way. They imagined the two strands of deoxyribose phosphate corkscrewing around each other in such a way that the backbone of each helix maintained a constant distance from the backbone of the other helix. The strands were sufficiently close for each base attached to one strand to make a loose contact with a base attached to the other strand. But because the distance between the helices was constant, the pairing between bases on each strand would have to be very specific: in the same way that all the steps in a spiral staircase must be the same width, the total width of a pair of bases, reaching across between one strand and another, would have to be the same. For this to happen, a purine (that is, adenine or guanine) on one strand could only pair with a pyrimidine (cytosine or thymine) on the other. A purine could not pair with another purine, because these larger molecules could not both fit into the space available – their combined width would be too great. And a pyrimidine could not pair with another pyrimidine, because their combined width would be insufficient to span the space between the strands.

The structure developed by Watson and Crick imposed an additional constraint that proved crucial: cytosine, one of the two pyrimidines, could only ever pair with one particular purine, guanine – never with the other purine, adenine. Conversely, the other pyrimidine, thymine, could only pair with the other purine, adenine, and never with guanine. The specific pairing proposed by Watson and Crick explained the otherwise baffling equivalence between the number of cytosine and guanine molecules in any sample of DNA, and why the number of thymines and adenines would also be the same.

But this model had another, much more far-reaching consequence. Because of the exacting requirements of base-pairing, in which adenine can pair only with thymine, and guanine only with cytosine, it is possible to predict with absolute certainty the order of bases attached to one of the two strands from the order of bases attached to the other. For example, a strand with the base sequence

ACCTTTGG

can only ever pair up with a sequence on a complementary strand that reads

TGGAAACC

The two would form a double helix can that can be summar-
ized as follows. I have not written in the details of the strands,
but the dashes represent the weak chemical bonds in the centre
of the double helix that join the bases from each strand together,
in the middle.

A – – T
C – – G
C – – G
T – – A
T – – A
T – – A
G – – C
G – – C

Because the chemical bonds joining the two helices are weak,
the two strands can easily be teased apart to make a complementary pair of single DNA strands:

A –   – T
C –   – G
C –   – G
T –   – A
T –   – A
T –   – A
G –   – C
G –   – C

Because each base can only pair with one other base, each of
these strands could be used as a template for the assembly
of a complementary strand from nucleotides in the general
environment:

```
A - - T   A - - T
C - - G   C - - G
C - - G   C - - G
T - - A   T - - A
T - - A   T - - A
T - - A   T - - A
G - - C   G - - C
G - - C   G - - C
```

Where there was one DNA double helix, there would be two, new double helices, each identical with the parent. If this sounds exactly like what chromosomes do during mitosis (cell division), then it should do – a chromosome is not made of DNA in some general way, but very precisely, from a single, double-helical molecule of DNA. During the process of mitosis the helices are teased apart, and complementary strands are created to produce two new double helices – and, therefore, two new chromosomes.

Watson and Crick had solved not only the structure of DNA, but had a theory of how the genetic information contained in DNA might be transmitted from generation to generation. The details of this process had, of course, still to be worked out: as Walter Gratzer notes in his anthology of writings from the journal *Nature*, the Watson–Crick structure was conjectural and had yet to find universal favour.[3] Indeed, doubts about DNA as the genetic material were still being voiced and published after the Watson–Crick model had appeared. Nevertheless, the significance was not lost on Watson and Crick themselves, who closed their landmark paper in *Nature* on 25 April 1953 with the notorious line, 'It has not escaped our notice that the specific pairing we have postulated immediately suggests a possible copying mechanism for the genetic material.'[4] The same issue of *Nature* contained papers by Wilkins, Franklin, and others, reporting the X-ray diffraction results that supported and informed the structure proposed by Watson and Crick. Franklin's comment, in her paper, that her data accorded with the Watson–Crick model should be no surprise in retrospect, as the Cambridge scientists had based their model, in part, on her data.

However, posterity rightly sees the Watson–Crick paper as a watershed in the development of biology. It is not for nothing that Walter Gratzer closes his *Nature* anthology with this paper, because it marks the transition between ancient history and the modern age.[5] The elegant, succinct structure of the genetic material, the way in which each of the bases could pair only with one other base, and the suggestion for how the information could therefore be faithfully copied and transmitted – all neatly answered the preoccupations of centuries. Watson and Crick's model showed definitively how both parents could contribute to the genetics of the offspring, and how this contribution might come in discrete units. Buffon's speculations about atomistic inheritance – revisited by Darwin and, ultimately, De Vries and the Mendelians – were thus shown to contain a large grain of truth. At the same time, the doctrine of preformation could also find some validation in DNA, in that the encapsulation of the essence of an organism in DNA makes sense, at least metaphorically, in terms of Bonnet's theory of 'essential parts'. Harvey might also have derived comfort from this result: a DNA double helix is hardly the same thing as an entire organism, but in its iconic use as the symbol of the origin and unity of all life, it makes a very good primordium. The slogan *Ex Ovo, Omnia* applies just as well in the age of DNA as it did in earlier times.

Implicit in the Watson–Crick model is a mechanism for how *changes* in the base sequence might be inherited. Any change in the base sequence of DNA is formally known as a *mutation*.

In his studies on the evening primrose that helped usher in Mendelian genetics, De Vries had coined the term to refer to a sudden, discontinuous change in the genetic material that would be expressed as a trait. Today we would use the word to mean a change in the DNA sequence, irrespective of whether it results in visible alteration. Most mutations have little or no effect, whereas some cause gross abnormalities to the structure or function of the organism.

The Watson–Crick model suggests that, at the most basic level, mutations are discrete alterations in the base sequence in which one base is substituted for another, or a base is gained or lost. Such misprints in DNA may change the 'meaning' of the gene, although this need not always be so. Mutations may be caused in all kinds of ways: some chemicals can react with and alter the bases in DNA. DNA also absorbs radiant energy – it is especially sensitive to ultraviolet light – and this can also damage the bases in a DNA chain. When Morgan and Payne exposed their fruit flies to X-rays and chemicals, they did not know precisely how genetic material might be affected by such mutagens because the nature of the genetic material was, at that time, obscure. However, they knew enough to suspect that X-rays and chemicals had the potential to introduce new variation, producing, say, the one white-eyed fly in a swarm of reds. Sometimes, though, mistakes occur in the DNA sequence during the copying process because, to put it simply, accidents do happen. The more often a piece of DNA is copied, the more opportunity there is for mistakes to occur – and accumulate.

Today it is very easy to take for granted how straightforwardly the Watson–Crick model explains the physical basis of mutation. It is worth remembering that, before 1953, there was no easy way to understand mutation, for there was no appreciation of its physical basis. Before Watson and Crick, only Buffon had come up with a coherent explanation, and that had been ridiculed for its invocation of a host of prior assumptions as well as for its plain want of physical, experimental validation.

The view of mutations as alterations in DNA sequence had another, profound consequence. With time, mutations would accumulate in DNA. More time would imply greater change, and because mutations are discrete, it is possible – at least in theory – to map the order in which mutations have occurred in DNA. In short, it is possible to trace the pattern of change in DNA as species evolve and diverge from one another. The birthrights of individuals, as of species, lie in the same DNA: changes in the microcosm will, with time, become reflected in the macrocosm. This idea would have resonated with nature-philosophy, which sought to reconcile the embryogeny of individuals with the position occupied by a species in the divine plan. The connection with Darwin's picture of evolution as a tree is almost too obvious to make. In short, the Watson–Crick model explains why the genealogies of people, flies and evolution as a whole look like trees, whose great boles and branches start as small twigs. As well as reflecting the pattern, the occurrence of mutation is now seen as absolutely vital for evolution as a process, for here, at last, is a way to generate the variation on which natural selection might act. Genetic mutation is the missing ingredient of the Darwinian model of evolution, and the Watson–Crick model suggested how it might occur. So much, indeed, was contained in that disingenuous line, 'It has not escaped our notice . . .'.

What the Watson–Crick model could not explain was why any particular base change should have any specific effect on the appearance or behaviour of the organism – nor, indeed, whether such effects might be mild or severe. Why should it be that some changes in base sequence turn brown eyes blue, whereas others create the equivalents of medieval monsters –

men with heads growing beneath their shoulders? In short, the Watson–Crick model said nothing about how genetic information was carried in DNA, nor how this information is processed to make itself visible – to give form to the formless. If we could read the information in DNA, what would we find? Would we see only yesterday's news, or the autobiographies of the archangels?

It was clear that the order of bases contained information that bore on the properties of living things. Because life had been associated with proteins since the 1830s, and because proteins were made of sequences of amino acids, the Watson–Crick model allowed people to imagine a line of bases coding for a line of amino acids, with a set of rules – a code – determining which bases 'stood for' which amino acids. However, the immediate problem was working out how just four bases might code for twenty amino acids. 'Tetranucleotide' models, in which combinations of four bases stood for a single amino acid, were popular, but in a startlingly ingenious piece of research published in 1961, Francis Crick and his colleagues proved that the genetic code was based on the number three.[6] That is, a 'triplet' of three bases would stand for single amino acids.

The simplicity of this result belies the thought that went into the experiments behind them, which can easily be taken for granted. It is not as if the researchers could have taken strings of bases and manipulated them directly, or rearranged strings of letters on a page. As with all genetics, the research was based on a careful choice of experimental subject. Where Mendel had peas and Morgan flies, Crick chose a virus: the so-called T4 bacteriophage (a bacterium-infecting virus), which infects *Escherichia coli*, a bacterium commonly found in the human intestine, where it lives, for the most part, so quietly that we go about our lives unaware that each of us carries many millions of these organisms. Sometimes a rogue strain appears: *E. coli*

146

strain O157:H7, for example, has been implicated in several food-poisoning outbreaks, but this is exceptional. Historically, the most famous strain of *E. coli* to be found in biochemistry and genetics laboratories carries the catalogue name K12. Sometimes the designation carries a mysterious suffix, K12λ (K12-lambda). The λ indicates that the bacterium carries an unwanted passenger, a virus called lambda, whose own genetic sequence has become part of the much larger genetic material of the bacterium.

Bacteria and viruses are far more different from one another than are mice from men, or indeed, flies from peas. Although they are very small, bacteria are complete cells, each with genetic material and a cell membrane. Some bacteria are parasites (and can cause disease), but most are capable of a free and independent existence. Viruses are very much smaller still, and consist of little more than their own genetic material wrapped up in a coat of proteins. Viruses are quite incapable of an independent existence and can propagate only by invading a cell and subverting the cell's biochemistry to manufacture more viruses. The infected cell eventually explodes, showering its environs with more infectious virus particles. Viruses attack all kinds of cells, but the first viruses to be studied were agents of disease in plants. Rosalind Franklin and her colleagues were justly famous for their studies on the structures of the protein coats of the tobacco mosaic virus.[7]

Given the difficulties of studying the genetics of peas, fruit flies, and so on, you might think that studying the genetics of bacteria and viruses would be quite impossible. Bacteria are so small that individual cells are only just discernible in conventional microscopes. And even when you can discern bacterial cells clearly, you find that one looks just like any other. So how are you to identify mutants, in the same way as an eye-colour mutant in a fruit fly? However, bacteria have diverse metabolisms, and can be distinguished by what they do rather than what they look like, so it is quite easy to select strains of bacteria by growing them on simple media which contain – or lack – certain nutrients. And although bacteria are individually visible only with difficulty, cultures of bacteria are easily visible to the naked eye as slimy spots or streaks. Mutant bacteria can be isolated from the patterns of growth formed by bacterial colonies growing on certain media.

Viruses are more difficult to study, because they can be cultured only in the cells or bodies of a host organism – without a host, viruses are as inert as house bricks. Bacteriophages are seen rather like a negative photographic image – from the empty, round holes of destruction they leave in bacterial cultures, the marks of infected bacteria which have exploded, spreading infection in all directions like blast waves. The genetics of bacteriophages are assessed by measuring the scale of such miniature epidemics. It seems incredible today, but Crick and his colleagues deduced the nature of the genetic code not in humans or fruit flies, nor even in *E. coli* – but in T4 bacteriophage, by careful observations of the circumstances in which various mutant viruses infected *E. coli*, and of the behaviour of the viruses that resulted from cross-infection (in which the genetic material of different strains of viruses could be crossed, almost as surely as Mendel crossing peas).

Even though Crick and his colleagues established that the genetic code was one in which a triplet of bases stood for a single amino acid, nobody knew which combinations of bases represented which amino acids. What was becoming clearer, however, was the mechanism whereby the information in DNA flowed to proteins, at least in a general way. Crick and other scientists began to conceive that information in a gene is 'read' in a two-step process. The first is called *transcription*, in which the DNA sequence in a gene is copied or, transcribed, into a complementary strand of another nucleic acid – ribonucleic acid, or RNA. This is very similar to DNA, except that the sugar in the sugar-phosphate backbone is ribose, rather than deoxyribose, and a base called uracil (U) is always found in the place occupied by thymine (T) in DNA. Uracil behaves exactly like thymine, including its specific pairing relationship with adenine (A).

For example, a sequence of DNA in a gene that looks like this:

CCCCAAGTAGGCGATTATTCTTATTCT

would be transcribed into a sequence of so-called messenger RNA that looks like this:

GGGGUUCAUCCGCUAAUAAGAAUAAGA

It is the sequence of triplets in this RNA – not the DNA – that constitutes the genetic code.

The second step is called *translation*, and occurs when the information in the strand of messenger RNA formed during transcription is translated into a sequence of amino acids. During translation, the messenger RNA is transported to one of a number of bodies in the cell called ribosomes. These entities, complex and fascinating in their own right, act as supports for the messenger RNA while its content is being read by a set of molecules called transfer RNAs. These molecules are like adaptors: at one end of each is a chemical prong made of three bases which can bind to a specific triplet of bases in the messenger RNA. The fact that transfer RNAs have this arrangement of three bases, rather than any other number, explains why the genetic code is arranged in triplets. The other end of each transfer RNA molecule is so formed that it binds to just one kind of amino acid. Sets of transfer RNA molecules, then, read the RNA bases in threes, using the information to construct, in the confines of the ribosomes, chains of amino acids.

The key, then, was in translation – to find which triplets of bases in RNA stood for which amino acids. This, the genetic code, was being worked out by others even as Crick and his colleagues were writing up their research on mutants in T4 bacteriophage. Towards the end of their account, ragged reportage broke through the polished surface of scientific urbanity, with a brief bulletin of late news: at a conference in Moscow being held as the paper was being written, an American biochemist, Marshall Nirenberg, had startled his audience with a report that he and a colleague, Heinrich Matthaei, had fed a substance called polyuridylic acid to a cell extract, and another substance called polyphenylalanine came out. Polyuridylic acid is a strand of RNA that consists only of uracils, and polyphenylalanine is a short stretch of a protein consisting only of the amino acid phenylalanine, molecules of which are strung together in a chain. The cell extract had 'read' polyuridylic acid as a strip of messenger RNA and had translated it into a string of amino acids. But the string was not random – it consisted only of phenylalanine. From this observation, Crick and his colleagues suggested that the amino acid phenylalanine was specifically encoded by a sequence of uracils in RNA, and that their own work suggested that it was a triplet of uracils that stood for one unit of phenylalanine.

Behind this simple idea lies an unknown complexity of biochemical interaction, but we can unpack it and study each step to make sense of what is going on. Imagine that you have a stretch of DNA that consists of adenines, like this:

AAAAAAAAAAAAAAAAAAAAAAAAAA

When this DNA is transcribed, the result will be a complementary strand of messenger RNA made of uracils, like this:

UUUUUUUUUUUUUUUUUUUUUUUU

This RNA is read by transfer RNAs, three bases at a time:

UUU UUU UUU UUU UUU UUU UUU UUU

A sequence of three uracils can be read only by the variety of transfer RNA that also recognizes the amino acid phenylalanine. Because of this, the stream of uracils is translated into a string of phenylalanines:

Phe–Phe–Phe–Phe–Phe–Phe–Phe–Phe

After Nirenberg and Matthaei's audacious experiment, it was only a matter of time before the genetic code was unlocked. The code varies slightly between organisms. However, in humans it is now known that of the sixty-four possible triplets that can be made from a palette of four bases, sixty-one encode one or other of the twenty common amino acids, and three are translated as 'stop' signs. The code is redundant in that a given amino acid might be represented by more than one RNA triplet. For example, the RNA triplet UUU is translated as 'phenylalanine'; but so is UUC, because phenylalanine can be encoded by either of these RNA triplets. The amino acid arginine is encoded by no fewer than six different RNA triplets (CGU, CGC, CGA, CGG, AGA and AGG) yet tryptophan has only one, UGG. One of the triplets in the genetic code has a double meaning. The RNA triplet AUG is rendered as the amino acid methionine. But this triplet also doubles as a signal for translation to start – the beginning of the gene. As a result, every protein begins with methionine, even if it is removed later on as the growing chain of amino acids is sculpted and folded to form the mature protein. However, nobody knows why some amino acids are represented by more triplets than others, or why the triplet for methionine, rather than any other amino acid, is also used as a 'start' signal. The answer must surely lie with the effects of time and chance on a genome, whose history stretches back billions of years.

Once the code had been understood, a mechanism for mutation came into view, in the sense that it was now possible to imagine how a change in base sequence might alter the appearance or behaviour of an organism. Some mutations have no effect at all. For example, altering UUC to UUU will still produce phenylalanine. Other mutations, however, accumulate in the genome and provide a rich store of information which can be used to trace differences between populations. Other mutations might change one amino acid to another. For example, changing AGG to UGG places a tryptophan in place of an arginine. In most cases, such changes either have no effect on a protein, or they contrive to alter its properties in some subtle way whose effects may or may not be immediately apparent. Changes like this underlie the differences between alleles of the same gene – explaining why, for example, the allele for tall peas differs from that for short peas. More than 1.5 million of these so-called single-nucleotide polymorphisms have been traced in the human genome, a vast library of tiny variations in gene sequences. However, by creating proteins with varying degrees of efficacy, these changes contribute to what we understand as the normal range of variation in healthy human beings. For example, it is now known that variations in the ability of people to smell faint odours, see colours of certain shades, respond to certain drugs or digest alcohol are all related to variations in the nucleotide sequences of genes.

Other mutations that alter the amino-acid sequence in a protein may have disastrous consequences, in that they produce a protein which does not work effectively – or doesn't work at all. Many inherited diseases, such as cystic fibrosis and Duchenne muscular dystrophy, have their origins in quite small misprints in DNA that can have very large effects. For example, inserting one extra nucleotide into a gene sequence will shift it out of register, so that the biochemical machinery that reads nucleotides in threes will start reading gibberish. Removing a nucleotide has the same effect. This is best understood by an analogy with this rather contrived string of letters:

DIDYOUSEEOURDADJIMANDHISBIGFATCATMAX

which can be made intelligible by breaking it up into threes, like this:

DID YOU SEE OUR DAD JIM AND HIS BIG FAT CAT MAX

However, the random insertion of an extra letter into the sequence –

DIDYOUSEEXOURDADJIMANDHISBIGFATCATMAX

creates nonsense if the string of letters is once again broken up into threes:

DID YOU SEE XOU RDA DJI MAN DHI SBI GFA TCA TMA X

In the world of genes and proteins, small insertions like this produce malformed, monstrous proteins which function poorly, if at all. From this we can see that mutations – in the sense of small changes to a sequence of nucleotides – have a natural history all their own, and the consequences of mutation may range from normal variation to egregious monstrosity.

However, understanding the connection between mutation and variation tells us relatively little about the part that genes play in many aspects of the human condition in which we are interested, such as our intelligence, aptitude for this subject or that, or other aspects of our personality. It seems clear that such things are to some extent inherited, but it may be hard, if not impossible, to implicate any particular gene. Here lies the crucial distinction that must be made between 'genes' and 'traits' that the early geneticists, in their zeal to avoid recognizing genes as real, failed to emphasize. Genes are individual units of inheritance, but traits are the external manifestations of the activities of an indeterminate number of genes. Whereas the trait of cystic fibrosis may rest on just one gene, this is exceptional – many traits have many genetic roots, and are also modulated to varying extents by environment or by lifestyle choices. However, the failure of the early geneticists to make this clear may have been what led to the popular misapprehension that there is a simple, one-to-one equivalence between genes and traits. For example, newspaper headlines often talk of the existence of genes 'for' diseases such as diabetes, schizophrenia or breast cancer, or personality traits such as criminality, homosexuality, aggression or good motherhood. But there exist no single genes for any of these traits, just as there exists no male-specific gene that causes men to build garden sheds or monopolize the TV remote control unit – and no 'melody gene' to explain the timeless musical talents of Johann Sebastian Bach, at least four of his sons and many of his relatives.

It is more likely that such complex, subtle traits stem from variations in the activities not of one gene working alone, but of hundreds or thousands of genes working together – in which case the role of any one, specific gene might be hard to understand in the context of the overall effect. In just this way, the generation of form from the formless might – like diabetes or schizophrenia or an urge to build a shed – be thought of as a single, complex trait which has many genetic roots. Given my operational definition of the genome as that single unitary entity that creates form from the formless, it follows that when considered as a trait, the creation of form must, by definition, involve every single gene in a single, operationally unified genome.

To understand the generation of form it is necessary to adopt a view of genetics in which genes are not important in themselves, but in how they behave with respect to all other genes in the genome. This is a rather new view of genetics – only in the past few years, as scientists have been able to discover the DNA sequences of whole genomes, have they begun to think of genetics in this way. But in itself, this 'network' view is not new. In fact, it has very deep roots indeed – roots that go back beyond Watson and Crick's model of DNA, back to the birth of genetics itself.

9

Operon

The first inklings that the interactions between genes were at least as important as genes themselves in framing the form and disposition of living things came as long ago as 1908, when an English physician named Archibald Garrod (1857–1936) noticed something curious about a collection of rare diseases in humans. The diseases were curious because they did not seem to be the result of infection or contagion. They did seem to occur more often in some families than in others, and this tendency was most marked when families were inbred. At a time when Mendelian genetics was still quite new, Garrod applied the lessons of Mendel's peas to humans, describing the inheritance of recessive traits – in this case, a tendency to contract certain rare diseases, which Garrod called 'inborn errors of metabolism'.[1]

One of these diseases is now known as alkaptonuria. Its symptoms include mottling of the skin and a progressive arthritis, and it afflicts approximately one person in every 200,000. Another disease in the same 'family' is phenylketonuria. This disease, which afflicts infants and young children, has a range of symptoms ranging from eczema to tremors and mental retardation.

Fortunately it is easily treatable, by strict adherence in childhood to a diet that does not contain the amino acid phenylalanine. One in 10,000 babies is born with phenylketonuria, making it one of the commoner inherited diseases. This frequency – and the need for sufferers to keep to a strict diet – explains otherwise mysterious slogans on jars and packets in supermarkets proclaiming that the contents 'contain a source of phenylalanine'. Garrod's insight was that these seemingly disparate diseases were the result of defects in inheritance, and that they were somehow connected. We now know that they are caused by inherited deficiencies in certain substances called enzymes.

Enzymes are examples of catalysts – substances that facilitate chemical reactions which might otherwise not occur at all. This constraint applies to virtually every active change within the body, from the growth of your hair and nails to the very first cell division of the embryo and all subsequent cell divisions; from the breakdown of food in your intestines to the exhalation of carbon dioxide in your breath. Enzymes are what builds your body and what breaks it down again. Without enzymes, the strands of DNA cannot be copied, nor can their cargo of genetic information be read or acted upon. But of what manner of substance are these everyday miracles that make life possible, these commonplace fixers of the marvellous? Enzymes – like the cells and bodies whose lives they control – are proteins, made of chains of amino acids. And herein lies a conundrum. If enzymes are part of the body and are responsible for its creation and function, then what makes the enzymes? This is where genetic information enters the story.

We now know that alkaptonuria is the result of a single genetic mutation, a misprint in the DNA. Normally, this piece of DNA – this gene – contains the instructions needed to make an enzyme called homogentisate 1,2-dioxygenase. The single task of this enzyme is to oversee the conversion of a substance

157

called homogentisic acid into another substance, maleyl ace-toacetate. Genetic mutation leads to an enzyme that doesn't work. Without it, homogentisic acid builds up in the tissues in the same way that rubbish piles up in the streets when the dustmen go on strike. Too much homogentisic acid, like too much rubbish, is a nuisance and can even be a hazard to health, and it is its accumulation that causes the clinical effects of alkaptonuria.

Phenylketonuria results from a different mutation, this time in a gene that contains the instruction for an enzyme called phenylalanine hydroxylase. This enzyme oversees the conversion of phenylalanine into another amino acid, tyrosine. A mutation in the gene for phenylalanine hydroxylase leads to a defective enzyme, unable to carry out this process. Phenylalanine builds up in the body until it becomes poisonous, causing the symptoms of phenylketonuria. This is why people with phenylketonuria should avoid eating foods that contain phenylalanine, or any other substance which the body might digest to produce phenylalanine.

Enzymes tend to have very specific functions. That is, each enzyme can catalyse only one very particular chemical reaction – there is no single enzyme that can break a molecule of phenylalanine into atoms all at once. The breakdown process is a cooperative venture, a kind of production line, in which each one of many small steps is governed by just one enzyme, in the way that each stage in a production line is overseen by a specialist worker or custom-built machine that can do that job, and no other – and yet all such workers or machines must be present for the production line to function. The instructions to make each enzyme are found in just one, unique gene – one gene for each enzyme.

Phenylalanine hydroxylase and homogentisate 1,2-dioxygenase are separate enzymes, but they both play their part in the process of breaking down the amino acid phenylalanine – it was this connection which, in retrospect, led Garrod to group alkaptonuria and phenylketonuria together in the same family of diseases. Phenylalanine hydroxylase, which catalyses the conversion of phenylalanine into tyrosine, is the first enzyme in the line. Homogentisate 1,2-dioxygenase – the deficiency of which leads to alkaptonuria – governs a step in the pathway much further down the chain, many steps removed from phenylalanine. In fact, this pathway is not a straight line but contains many branching points, and products from other pathways may join it. The phenylalanine breakdown process represents a network of cooperative genes, working in the context of the genome as a whole. And yet, at any point in the network a genetic mutation can disable one of these enzymes, causing disease. Because mutations are carried in the genes, these diseases are inherited.

A failure of any one of these enzymes, at any point in this process, can result in the pathological accumulation of one by-product or another, and it is this that causes disease. Each enzyme is therefore associated with its own inherited disease – but because all the enzymes are concerned with the same pathway, these diseases may share certain features. For example, people suffering from one or other of this family of related diseases may have fairer complexions than might be expected from their family histories, a result of a deficiency of skin pigmentation. Conditions such as albinism – in which the skin contains no pigment at all – are caused by defects in the metabolism of the amino acid tyrosine. Because tyrosine is an important by-product of the digestion of phenylalanine, people with phenylketonuria or other closely connected diseases may also seem unusually blond.

Garrod lived in a time before genes were understood in terms of chromosomes and DNA, and so would not have understood

the problem in terms of genes and enzymes. The modern concept of the gene as a unit of inheritance came with experiments
by two American scientists, George W. Beadle (1903–89) and
Edward L. Tatum (1909–75). After gaining his doctorate in
1931, Beadle went to do research on fruit flies with Morgan.
His work in Morgan's lab and afterwards led him to suspect
that the variation of traits in fruit flies might not be the result
of simple mutations in genetic material, but the summation of
entire chains of events at the biochemical level which were
controlled by networks of genes acting together. The link
between gene and trait was not a simple one-to-one correspondence, but the result of a long and sometimes circuitous path
of biochemical events, rather like the events which we now
know result in Garrod's family of diseases.

Beadle's career took him to Stanford in California, where he
started to work with Tatum on an entirely different organism:
a species of fungus, the bread mould *Neurospora crassa*. From
breeding experiments involving a range of mutant mould strains
unable to digest one compound or another, the scientists were
able to map *metabolic pathways* – similar to the one which,
in humans, leads by incremental steps from phenylalanine to
homogentisic acid and beyond, with each step controlled by a
single enzyme. It was Beadle and Tatum who came up with
the theory that might be written on a T-shirt: 'One Gene, One
Enzyme'. This model means exactly what it says – that for each
enzyme in a metabolic pathway, there had to be just one gene.
The genome is a list of instructions to make a pharmacopoeia
of enzymes, and from these enzymes bodies can be built.[2]
Importantly, Beadle and Tatum showed that traits were not, as
a rule, switched on and off by single enzymes, but were – as
Beadle had begun to suspect in his days with Morgan – the
end states of a whole pathway of enzymes, the structure of each
one of which was determined by a gene. What we see in

organisms is the result of networks of genes working together. This message has, unfortunately, become overshadowed and obscured by the catchy 'One Gene, One Enzyme' slogan, which is all too easily elided into 'One Gene, One Trait'.

The birth of the 'network' view of genetic activity as we now understand it came in 1961, with the publication of an article by two French researchers, François Jacob (b. 1920) and Jacques Monod (1910–76). The article was entitled 'Genetic regulatory mechanisms in the synthesis of proteins' and contained a new and significant word – 'operon'.[3] Jacob and Monod worked mostly at the Pasteur Institute in Paris, and began their collaboration in 1958. Among their many accomplishments was the proposal for the existence of messenger RNA, and also the discovery of so-called regulatory genes, whose function was not to produce an enzyme but to control the activities of those that did. Such entities are essential elements of control in a network of interacting genes.

Where Crick and his colleagues worked out the nature of the genetic code by studying the viruses that infested the gut bacteria E. coli, Jacob and Monod were interested in the bacteria themselves. Their particular concern lay in understanding how the bacteria digested the sugars on which they fed. One of these sugars was lactose, the sugar commonly found in milk. In chemical terms, a molecule of lactose is two molecules of a simpler sugar, glucose, stuck together. Lactose, on its own, is of no use unless it can be broken down into glucose – a common energy currency for all forms of metabolism, whether bacteria or people. E. coli bacteria split lactose with the help of an enzyme called beta-galactosidase.

Jacob and Monod noted that bacteria did not produce this enzyme all the time, but only when necessary – that is, if there were any lactose around to digest. This observation, in itself, was not news. Scientists had known for some years of enzymes which appeared only when they had a job to do. Where Jacob and Monod broke new ground was in how they approached the phenomenon, in terms of the regulation of genes. Perhaps, they thought, there were two kinds of gene. The first would be the 'structural' genes, coding for enzymes and other proteins we can readily see and measure. Then there would be a more shadowy class of genes – the regulatory genes, which would oversee the activities of other genes, ensuring that they were 'switched on' only when needed. The proteins produced by regulatory genes would be far less abundant than those made by structural genes, and consequently much more difficult to isolate.

Beta-galactosidase, like all enzymes, is created from information held in a gene. This gene is transcribed into messenger RNA and translated into the string of amino acids that make up the finished protein enzyme. But if the enzyme is made only when there is lactose to digest, Jacob and Monod speculated, there must be a mechanism for detecting the presence or absence of lactose and switching the beta-galactosidase gene on or off, as necessary. In what is now seen as one of the classic experiments of modern genetics, Jacob and Monod showed that structural genes, such as the one containing the code for making beta-galactosidase, were ordinarily switched off, or 'silent', unless prompted into action by the appropriate environmental stimulus, in this case the presence of lactose. 'Silence' in this context means that the beta-galactosidase gene was not being transcribed into messenger RNA, so no enzyme would be made. Jacob and Monod reasoned that genes would not be silent of their own accord, but would have to be silenced by some external agency. However, this selfsame agency would also loosen its hold in the presence of the appropriate stimulus, such as a lactose molecule. There had to be something that acted as both sensor and censor.

It quickly became apparent that this molecular censor had to be quite distinct from beta-galactosidase. Jacob and Monod showed that strains of *E. coli* existed which, by virtue of mutations in the beta-galactosidase gene, were unable to digest lactose by themselves, and had to be provided with glucose by the researchers, else the unfortunate microbes starved to death. But Jacob and Monod found another, more interesting kind of mutant strain in which the bacteria produced perfectly normal beta-galactosidase – but did this whether lactose was present or not. This indiscriminate synthesis of the enzyme could not have been connected with any damage to the beta-galactosidase gene, because the enzyme itself was quite normal, so something else must have been happening. Jacob and Monod supposed that some kind of mutation was disabling a gene at one remove – a gene that would normally produce a substance whose sole job it was to prevent the transcription of the beta-galactosidase gene. Inactivation of this blocking gene would lead to the constant transcription of the beta-galactosidase gene, however much lactose was present in the environment. The product of this blocking gene (which was later identified as a protein) was called a 'repressor'. The gene encoding the repressor was not a structural gene, but a regulatory one: a gene whose function was to control the activities of other genes.

Jacob and Monod identified yet a third kind of mutant connected with the digestion of lactose. There exist mutant bacteria that have normal genes for both the repressor and for beta-galactosidase, but *still* synthesize the enzyme irrespective of the presence of lactose. It turns out that these bacteria carry mutations in an untranscribed section of DNA next to the beta-galactosidase gene. This section of DNA became known as the operator. For the repressor protein to silence the beta-galactosidase gene, it first has to attach itself to the operator. Mutating the operator is a bit like bombing a runway so that planes can no longer land there. The repressor protein might be normal, but if the operator sequence is so disfigured by mutation that the repressor cannot bind to it, then the production of beta-galactosidase will carry on as if the repressor weren't there at all.

By combining all these details, Jacob and Monod painted the first picture of what we now see as the network view of genetics, a simple case of genetic regulation in which bacteria would be able to produce the enzymes they needed in order to digest a simple sugar, and to do so only when they needed it. When there was no detectable lactose in the environment (which is the case for most of the time) the repressor gene ensured that there was always enough repressor protein around to sit on the operator, so that the beta-galactosidase gene was switched off and no beta-galactosidase made. But when lactose was present, molecules of lactose mobbed the repressor, preventing it from attaching itself efficiently to the operator. The beta-galactosidase gene was then free to produce enough enzyme to digest the lactose. When the job was done, and lactose fell below the concentration required to interfere with the repressor, the repressor would take up residence once more on the operator, and the beta-galactosidase gene would be switched off again.

The genes for the repressor, for the operator and for beta-galactosidase were thus united by their function: to digest lactose, when the opportunity arose. But Jacob and Monod found something else – that all three entities sat close together on the same strand of DNA. They saw the significance of a cluster of genes associated by location as well as function, and called such a cluster an operon. The lactose-digesting operon became known as the *lac* operon.

Many other operons have since been found in bacteria of all kinds, each one containing a set of structural genes necessary to perform a certain function, and regulatory genes to ensure that the structural genes are switched on only when necessary. The *lac* operon is very simple – just one structural gene, one regulatory gene and an operator – but the significance of the concept of the operon goes well beyond the realm of bacteria. What is a genome if not a kind of operon, a physically associated cluster of genes, all of which share a single task, the creation and maintenance of an organism? If the genome is an operon whose function is to create and maintain an organism, it must contain genes directly necessary for that function: structural genes to produce enzymes as diverse as, say, phenylalanine hydroxylase and beta-galactosidase, as well as the proteins such as collagen and keratin from which bodies are made. But the genome must also contain regulatory genes that control the activities of these structural genes, to ensure that they are switched on at the appropriate times and in the correct sequence, so that the result of the sequence of cell divisions at the start of the development of an individual gives rise to an embryo, functioning and complete in all its parts.

The genome of any organism is conceptually equivalent to the *lac* operon of *E. coli*, if far more elaborate. Jacob and Monod's epochal work made the point that the path between DNA and the organism lay not in the information that DNA contained, but in how that information was controlled. The significance of the work was not lost on the scientific community, and Jacob, Monod and their colleague André Lwoff (1902–94) were awarded a Nobel prize in 1965.

Jacob and Monod did not actually succeed in isolating the *lac* repressor. It is in the nature of regulators that only a few molecules are required at any one time for them to be effective, so isolating a regulatory protein is rather like searching for a particular strand of microscopic hay in a haystack itself far too small to see with the naked eye. The task, however, was accomplished in 1967, by a scientist named Walter Gilbert, who went

165

on to pioneer the DNA sequencing technology eventually used to sequence the human genome. Gilbert won a Nobel prize in 1980 for his work on genetic sequencing. Meanwhile, also in 1967, a young academic called Mark Ptashne isolated another repressor, an achievement which earned him a full professorship at Harvard at the relatively tender age of thirty-one. Ptashne's work underscores the point that a genome is really an operon writ large, for the repressor he isolated was responsible for the control of an entire genome – that of a virus named bacteriophage lambda.[4]

Like all viruses, lambda consists of an inert string of genetic code, packaged for safety in a protein coat. When a virus meets a bacterium, the protein coat sticks to the bacterial cell wall, but the viral genome makes its way inside the cell itself. There, the genes are read by the bacterium's own transcription and translation machinery, churning out more copies of the viral genome and the proteins that comprise the protein coat. Soon the entire effort of the bacterium is converted to the manufacture of viruses. Eventually the bacterium, bloated with viruses, explodes – scattering thousands of new viruses throughout the surrounding medium. Some of the viruses will meet other bacteria, promoting the infection.

And so it happens, but only some of the time. For lambda has a secret life. Sometimes a lone virus will infect a cell and, rather than furthering the spread of infection, will splice its genome into the much larger genome of the bacterium. Once inserted, the viral genome will behave like any other piece of bacterial DNA. When the DNA is copied before cell division, the viral DNA will be copied too. The genome of the virus can remain, silent within the genome of its host, for many generations. But, just now and then, a sleeping virus – a distant descendant of the original infectious particle, in a host equally remote from the first victim – will cut loose. Lambda is not alone in this habit. Some viruses that cause disease in humans, such as the herpes simplex virus, can hide out for decades in exactly this way, silent within the genome of its host.

But how can a virus stay silent at all, given the natural tendency of viruses to hijack cells immediately on infection? Some viruses, such as lambda, clearly have a choice – destroy a host immediately, or hide out and defer destruction to another day. And as in the case of the 'decision' of the *lac* operon to synthesize an enzyme or remain quiet, the mechanism of choice is governed by a repressor.

In the 1950s it was found that some strains of lambda were always destructive, and never established a dormant state. The cause was reasoned to be a mutation in a regulatory gene, the so-called lambda repressor gene. This gene would contain the code for a protein that sits on defined parts of DNA – operators – so that the transcription enzymes are denied access to the genes. The lambda repressor would function in a very similar way to the *lac* repressor. However, rather than working within just one operon in a genome, the lambda repressor would block transcription in the entire lambda genome, a collection of some fifty genes. In this sense, the whole lambda genome could be thought of as a single operon. Mutant viruses, unable to produce their own repressor, would always be switched on, rather like a *lac* operon unable to make a *lac* repressor. But when viruses hide in a genome, all their genes would be switched off except one – the one that produces the repressor itself. This strategy would have an additional advantage for the virus that gets in first. Because there is a low level of repressor protein in an infected cell, the cell is protected from further infection by viruses.

All this begs the question of how a dormant virus might pick its moment to wake. The answer lies in the mechanism of mutation itself. Ultraviolet light is a potent mutagen, in that DNA is especially sensitive to it. This explains the link between sunbathing and skin cancer: tanning is the body's natural response to UV as it tries to shield itself from the Sun's rays. Our skin cells have an array of enzymes that repair DNA damaged by UV light. Mutations in one or other of these enzymes leads to a variety of syndromes in which patients are more than usually susceptible to skin cancer. Many bacteria are rather poor at repairing DNA damaged by UV light, and *E. coli* – which naturally spends its life in the human intestine, a place where, proverbially, the Sun never shines – is one of them.[5] UV light is as lethal to a virus lying dormant in a bacterial genome as it is to its host, unless the virus can escape before the host dies. To do so, it must find a way to activate its genome and make more virus particles, but this cannot happen unless it can find a way to inactivate its own repressor, liberating its own genes from enforced silence. But what could repress the repressor? It turns out that the physical environment plays its part in the process: UV radiation is especially damaging to DNA, and the lambda repressor gene is especially sensitive to it. Degraded by UV, the repressor gene is no longer able to keep synthesizing repressor molecules. Released from bondage, the genes in the lambda genome awake from their long dormancy, the viral genes are transcribed and the long-delayed infection can continue.

All this remained rather speculative, in the absence of physical evidence for the repressor. It fell to Ptashne and his colleagues to hunt down this hitherto elusive creature, which they did. They crowned their remarkable achievement by catching the repressor in the course of binding to the operator, isolating it, crystallizing it and subjecting it to X-ray diffraction studies of the kind pioneered by the Braggs and Franklin. In this way, Ptashne's team obtained a picture, in atom-by-atom detail, of how a repressor protein actually interacts with an operator – genetic regulation at work.

The *lac* operon and bacteriophage lambda are simple examples of genetic switches. With the *lac* operon, the repressor has the relatively mundane task of controlling a single gene – the gene for an enzyme, beta-galactosidase – ensuring that it is only produced when it is required. The scope of the lambda repressor is much greater, for it controls all the genes in a genome. The genome is very simple, to be sure, but the point is made: the control exercised by the shepherd-like lambda repressor on the flock of its genome shapes the destiny of the entire organism, and that is surely what a genome is all about.

The suspicions of Garrod, reinforced by the careful work of Beadle and Tatum on the mould *Neurospora*, showed that there was more to genetics than birds on a wire. To posit a direct, one-to-one correspondence between genes and traits was a caricature of the real thing. What we see as disease states, or the range of normal development, reflects the activities of many genes – perhaps all the genes in a genome – working together in a harmonious way. In this holistic model, in which genes interact as a network, the precise role of any one gene can be hard to tell. As Jacob and Monod so elegantly showed, mutations in any one of the structural gene, the operator or the repressor gene could explain why *E. coli* bacteria produced too much beta-galactosidase, or none at all: but only when all three genes are considered together could their function be fully understood.

Jacob and Monod's work on the *lac* operon, dramatically extended by Ptashne on the lambda repressor, was a harbinger of greater things to come, and a view – emerging only now, with the relatively easy, large-scale sequencing of genomes – that the function of individual genes is less important than how the various functions of genes interact in a network in which, behind the visible storefronts of structural genes, regulatory genes are hard at work. Even though regulatory genes are even harder to grasp, as physical realities, than are genes responsible for enzymes or proteins such as collagen, the facts of genetic regulation are all around us, waiting to be investigated. The world is shaped by genetic regulation. Because of such regulation, the aphids studied by Bonnet, in which generation succeeds wingless generation, can suddenly grow wings and fly away when their bodies sense the presence of ladybirds. Thanks to genetic regulation, a hydra when cut in two can regenerate the missing parts of its body, the phenomenon of regeneration that so captivated Trembley and Haller, and was the subject of the work of Morgan's youth and old age. Thanks to gene regulation, plants can appear with leaves instead of petals, suggesting to Goethe that laws of form might exist beneath the variety of life.

Thanks to genetic regulation, a single, spherical cell can divide, divide again and – within 28 days – become a recognizable miniature of a human being. Creating a human embryo, however, implies a degree of regulation far more sophisticated than the once-only events that turn single genes on or off, as in the *lac* operon. But this is a relatively simple thing, the first case of the phenomenon of genetic regulation to be described – and, as such, an exemplar, a proof of principle and an icon. But there is no reason why regulation need be limited to a single event, or that a single regulator should act only on structural genes. Regulators can interact with other regulators, producing a cascade of regulation – offering untold opportunities for subtlety and flexibility.

Even then, the secret of genetic regulation is the ability to respond to changing conditions, be it the presence or absence of lactose, UV radiation or ladybirds. Arguably the most changeable environment in the whole of nature is the developing embryo. The regulatory genes that direct the division of a fertilized egg into two cells, within minutes of fertilization, and two into four within hours, find themselves in a completely different environment with each division. Groups of cells require the activation of whole batteries of structural genes which would be quite superfluous in a single cell: genes, for example, coding for proteins that keep two otherwise separate cells stuck together so that they can become a tissue; or proteins that make up cell-surface receptors, and actively controlled pores that allow the passage of some substances – but not others – through cell membranes, so that cells can communicate with one another.

It is thanks to regulatory genes that these developing tissues can interact to create yet further tissues and organs. In the very early embryo, interactions between the ectoderm and the endoderm create an entirely new layer of cells, the mesoderm. Further interactions shape the mesoderm into the notochord, the tissues that will become the somites, and the lateral plate mesoderm that will become the body wall. The notochord, once formed, coerces the neural tube into closure; interaction between the somitic and lateral plate mesoderm creates the kidneys. Meanwhile, regulatory substances secreted by the primordial germ cells, returning from their long journey to the yolk sac, sculpt parts of the body wall into the sex organs; regulatory substances found in males – but not females – ensure that these sex organs become testes, and the primary germ cells become sperm.

Couched in these terms, the development of an individual – the creation of form from a formless egg – is a dynamic trait, a consequence of a cascade of genetic, regulatory interactions, no different in principle from the sequences of enzyme-controlled metabolic events that Beadle suspected lay behind the visible expression of each trait in flies and, later, *Neurospora*. Searching for the regulatory network that creates the human embryo, or indeed the embryo of any complex organism, poses a greater challenge than the investigation of mould metabolism.

For one thing, mutations in regulatory genes which are important in the network that creates an embryo are unlikely to produce creatures with small but picturesque mutations – such as the single white-eyed fly among reds, or a dwarf pea plant where all others are tall – which represent rather small quirks in the normal range of variation, and which can be used in breeding experiments. Such changes are small ripples on the surface of a deep and murky pool of regulation – minor alterations made once the structure of the organism is all but complete. On the contrary, mutations in regulatory genes will produce things altogether more puzzling – those same marvels that fascinated Paracelsus, Paré and Bacon, captivated Geoffroy, and drove Bateson to compose medieval bestiaries recording insects with legs growing out of their heads. In short, monsters.

10

Monsters reloaded

Nursery rhymes can be unspeakably violent. Recall, for example, the three unfortunate mice who, as well as being blind, had their tails docked by a sadistic, knife-wielding peasant. The real–life, adults–only version of this tale features a newborn mouse, seemingly perfect in every way except that the front end of its body is oddly truncated, ending in a small mound crowned by two tiny ears. This mouse is certainly blind, since not only does it have no eyes, it also has no head. It looks as cleanly decapitated as if by the farmer's wife of folklore. The mouse was one of four born without a head, as a result of a mutation in a regulatory gene called *Lim1*. With no mouth or nose to breathe through, the mouth died very soon after its birth. More than a hundred other headless mice foetuses did not get as far as being born.

The experiments in which these mice were created[1] were part of an effort to understand the activities of *Lim1*, one of an increasingly well-documented cadre of regulatory genes whose role it is to ensure that every part of the body develops in the place it should. When such genes are mutated, the result can be as monstrous as anything from nursery folklore. The work on the headless mice is just one of many examples of studies of genes whose action goes well beyond the maintenance of what we would consider the normal range of variation. The genetic variation that dictates whether your eyes are brown or blue is superficial – but there is a more fundamental kind of variation, in genes that determine whether you have a head at all for eyes to stare from.

Lim1 belongs to a group of regulatory genes called *Hox* genes, and their story really begins with Bateson's descriptions, in *Materials for the Study of Variation*, of insects with the *antennapedia* mutation in which tiny simulacra of legs grew out of the head in place of antennae; of insects with extra wings; and of people in which vertebrae in one part of the spinal column had been transformed into vertebrae characteristic of an adjacent region of the spine. Bateson realized that all these syndromes were characterized by a kind of abrupt discontinuity in which body parts developed in places that would normally be occupied by others, but with the rest of the body eerily undisturbed. He called this phenomenon homeosis, and the genes responsible for such mutations became known as homeotic genes.

Decades later, a family of homeotic genes were each found to have a very characteristic region of their DNA, 180 bases long, which, when translated into the amino-acid sequence of a protein, made a kind of prong which fitted neatly onto the operator regions of other genes. When geneticists wish to pinpoint specific parts of a DNA sequence on a computer printout, they often draw a rectangle or 'box' around it. The DNA-binding region of homeotic genes therefore became known as the 'homeo-box',[2] later contracted to *Hox*. Other families of genes have since been discovered, each with their own distinctive DNA-binding sequences. One such gene was found to be responsible for a mutation in *Drosophila* known as *paired*, in which the segments of the embryo failed to develop properly.

This gene contains a characteristic sequence known as the *paired* box, and a family of related genes has since been found in many other animals.[3] This family has acquired the name of *Pax* genes, and its most famous member – *Pax6* – is involved in the development of eyes throughout the animal kingdom. Mutations in *Pax6* turn out to be implicated in congenital eye defects in humans as well as in mice and flies, and the artificial expression of *Pax6* in flies outside the head has created bizarre mutant flies in which eye tissue appears all over the body. *Pax6* demonstrates a startling coherence of plan which unites otherwise diverse animals. (One cannot help feeling that Goethe and Geoffroy would have approved.) But whether they are called *Hox* or *Pax*, such families of genes clearly encode proteins similar in concept to the *lac* or lambda repressors – that is, they are regulators.

Bateson advanced homeosis as just that kind of discontinuous variation that might stand in opposition to the 'insensible gradation' of Darwinian change. He thought that such clear-cut discontinuities were a feature of all genetic variation. In that, he was mistaken. However, what he had discovered was a stratum of genetic variation beneath the kind of variation studied by the biometricians – a genetic substructure over which the everyday kind of normal variation is draped, and on which it depends for its expression.

Such variation is, almost by definition, hidden, and very hard to reach in conventional breeding experiments. The shock value of monsters is a function of their rarity: because monstrosities involve major disruption to body form, it is rare that they live long enough to be born, let alone be capable of reaching adulthood and taking part in the kinds of breeding experiment directed by Mendel or Morgan. Headless mice can be born only because of careful breeding of a stock of mice, each of which carries one recessive, mutant allele of the *Lim1* gene.

Such a mouse will appear normal because the decapitating effects of the recessive, mutant allele will be masked by the dominant, normal allele on the accompanying chromosome – one normal allele out of the pair is all that is required to get a head. Only when these apparently normal mice are interbred is there any chance of producing headless offspring. Because the headless form of the mouse depends on the conjunction of two recessive *Lim1* alleles, potentially headless embryos will be produced in the classic ratio of three normal offspring for every one that is headless, just like Mendel's dwarf pea plants. In the case of the headless mice, however – unlike the dwarf peas – very few survive long enough to be born and have their condition reported.

The rarity of recessive monsters was just one of the problems that confronted Edward B. Lewis, a pioneer of research into what became known as the *Hox* family of genes, who started in Morgan's footsteps by breeding flies – but with the ambitious aim of mapping the genes that, when mutated, created the whole homeotic freak show. Lewis joined Caltech's successor to Columbia's Fly Room, obtaining his doctorate there in 1942, and was especially interested in an extravagant homeotic mutation called *bithorax*. The middle section of the fly, behind the head and in front of the characteristically limbless abdomen, is called the thorax. This is the fly's engine room, carrying everything it needs to move from place to place. Each of the three thoracic segments bears a pair of legs, and the second segment a single pair of wings. In insect terms, flies are highly evolved – more primitive insects, such as dragonflies, have two pairs of wings. In flies, the second pair of wings – sprouting from the third thoracic segment – has evolved into a pair of drumstick-shaped structures called halteres. These are organs of balance, which whirl around like gyroscopes and help make flies such expert aeronauts. Flies with the *bithorax* mutation are hard to miss because they have not one but two pairs of wings: one pair growing from the second segment, as normal, and another emerging from the third. A closer look reveals that this mutation is more than just the addition of wings. The entire third thoracic segment, rather than developing as normal, has developed as a duplicate of the second segment.

With immense care and decades of patience borne out of thousands of breeding experiments, Lewis found that the gene responsible (named *bithorax*, after the mutation resulting from defects in the gene) was part of a closely linked cluster of homeotic genes involved in specifying the structure of the thorax and abdomen of the fly. This cluster was similar to an operon, except that *all* the genes involved were regulators. Indeed, some of them appeared to regulate themselves as well as one another, offering the possibility of great control over detailed structure. Another cluster of homeotic genes was found to be involved in the structure of the head end of the fly. This cluster was the 'ANT-P' or *antennapedia* complex, containing the *antennapedia* gene responsible for the monsters that Bateson had noticed and catalogued, those insects with legs growing from their head in place of antennae.

It turned out that the fly is unusual in that what was once a single cluster of genes has been split into two: the *antennapedia* and *bithorax* complexes. This may be a peculiarity of flies and their close relatives. In most animals more complex than jelly-fish, there is single cluster – the *Hox* cluster – which is equivalent to the two clusters in flies. This cluster can be thought of as a single operational unit whose members are responsible for ensuring that each segment of the animal develops the appropriate structures. Failure to do this results in homeosis, in which structures are matched with the wrong segments – legs on the head, for example.

Lewis discovered a particularly striking feature of the *Hox* cluster: the genes are arranged within the cluster in the same linear order as the segments in the animal whose identity they govern. For example, the *Hox* genes responsible for the head end sit next to the genes responsible for the thorax, and further along the cluster are genes responsible for setting the identity of progressively posterior parts of the abdomen.[4] This remarkable feature, known as collinearity, can hardly be a coincidence. The *Hox* genes are of such importance in creating the shape of an animal that close and coordinated regulation is essential. As many *Hox* genes regulate one another, it makes intuitive sense for a gene involved in specifying a particular part of the body to be more concerned with genes involved with the formation of adjacent regions, rather than with ones farther away.

This close regulation presumably explains why *Hox* clusters have been found in all animals with a distinct front and rear end. Animals as diverse as mice, humans, flies and worms have at least one cluster of *Hox* genes in which the genes are arranged in an order corresponding to the position of the structures they specify. Simple animals such as jellyfishes and *Hydra* (of regenerative notoriety) also appear to have *Hox* genes arranged in a linear order. Sponges, the simplest multicellular animals, appear to have isolated *Hox* genes, but not, it seems, a linear cluster of them – and this may be connected with their lack of any regular body patterning above the level of simple tissues. Single-celled creatures such as yeasts also have *Hox* genes, although, as in sponges, they are not in clusters. Plants do not appear to have *Hox* genes closely related to those found in animals, probably because plants do not organize their bodies in the same stereotyped and tightly organized way that is characteristic of animals. Plants do, however, have other homeotic genes, mutations of which lead to the kinds of monstrosity – petals developing as leaves, and so on – that led Goethe to suggest that plant structures have a unity of plan in which all organs can be seen as elaborations on an archetype of the leaf.

The closeness with which *Hox* genes regulate one another is illustrated – startlingly – by the fact that *Hox* genes from one species can regulate corresponding *Hox* genes in another, totally different species. A *Hox* gene called *deformed*, for example, is involved in shaping the back of the head in flies. In addition to the DNA sequence that encodes a regulatory protein, the *deformed* gene contains a regulatory element akin to the operator sequence that Jacob and Monod found in the *lac* operon. This element is essential for the *deformed* gene to do its job properly. Mammals, such as mice and humans, also contain a gene very like *deformed* which is involved in the development of the back of the head, specifically part of the hindbrain. It is reasonable to suppose that the mammalian and fly versions of *deformed* are homologues – that they share a common evolutionary descent from a gene which shaped the backs of the heads of the long-extinct common ancestor of flies and mammals some 600 million years ago.

Given that all other traces of this common ancestor vanished more than half a billion years ago, it is remarkable that genes from flies and mammals can stand in for each other. In 1992, Alexander Awgulewitsch and Donna Jacobs of the Medical University of South Carolina took the *deformed* regulatory element from flies and showed that it regulated the activity of the mouse version of *deformed*, in the mouse hindbrain. To show that this was no fluke, William McGinnis of Yale University and colleagues did a converse experiment, showing how regulatory elements associated with the version of *deformed* found in humans could substitute for the *deformed* regulatory element in embryonic flies.[5] These amazing results show that *Hox* genes have been as vital to the evolution of animal form as they are to the creation of every new animal embryo – whether fly maggot or human baby.

Had a messenger arrived at the Académie des Sciences in the spring of 1830, bearing a message from the future pointing out these remarkable similarities in the genetic substructures of animals, Geoffroy would have cheered, Cuvier would have gnashed his teeth in impotent rage, and Goethe would have smiled in quiet satisfaction. It does beg the question of why, given these similarities, animals come to look as different as they do. *Hox* genes or not, maggots remain maggots, babies are still babies, and Jeff Goldblum will never really make that Kafkaesque metamorphosis into a giant insect.

Some of the differences are related to the numbers of *Hox* clusters in a given genome. Although most animals have just one *Hox* cluster, the genome of vertebrates has been duplicated at least once, leading to the wholesale duplication of *Hox* genes. So whereas flies have a single *Hox* cluster, mammals such as mice and humans have four. Although a *Hox* gene in one cluster may correspond to genes in the other three clusters, the duplication has created the possibility that each new gene can evolve in its own way, giving immense scope for greater complexity and refinement in gene regulation. It is possible that the great structural complexity of vertebrates compared with other animals results, at least in part, from this multiplicity of *Hox* genes.

In 1994 this possibility was thrown into sharp relief by the discovery, by the British zoologist Peter Holland and the Spanish researcher Jordi Garcia-Fernandez, that lancelets have a single *Hox* cluster.[6] Given that these vaguely fish-like animals are the closest relatives of the vertebrates, but lack quintessentially vertebrate attributes such as the neural crest, it is possible that we owe our brains, skulls, teeth, sense organs, limbs – indeed, everything in which neural crest tissue plays a part – to a duplication of a primordial *Hox* cluster in our fish-like ancestors more than 550 million years ago.

But lest we are tempted to use *Hox* cluster duplication as a way to justify our self-appointed station at the pinnacle of creation, we should note that some vertebrates have gone even further. The most diverse and successful group of vertebrates by far is the teleost fishes, encompassing virtually all familiar species of bony fishes (and many unfamiliar ones). There are more species of teleost than of all other species of vertebrate – other fishes, birds, amphibians and mammals – put together, and they are found in a great diversity of form, ranging from tuna to seahorses. The reason for this great diversity and pronounced success may lie in *Hox* cluster duplication: some teleosts appear to have no fewer than eight *Hox* clusters. Even the humble zebra fish (*Danio rerio*), stalwart of aquariums in the waiting rooms of a million dentists, is known to have seven *Hox* clusters.[7]

Other large-scale differences in animal form are undoubtedly accounted for by other differences in gene structure, less obviously dramatic than bulk duplication. Although *Hox* genes in a fly are homologous with those in a human, they differ in a host of details as regards sequences and structures. For example, the *Drosophila* cluster has relatively small numbers of rather extensive genes, whereas each human *Hox* cluster contains many more genes, but physically smaller – an entire human *Hox* cluster could fit into the space occupied (in terms of bases of DNA) by *antennapedia*, just one of the genes in the *Hox* cluster in fruit flies. Such differences, accumulated over history, might ultimately contribute to differences in the physical appearances of animals.

In any case, humans and flies are so divergent, in terms of their evolutionary histories, that it is hard to be certain about how they got that way, relative to each other. Comparisons between more closely related creatures reveals a clearer answer – that animals differ in appearance not so much because of the accumulation of mutations in structural genes (although that is surely important), but because of evolutionary changes in the way that genes are regulated. So, while flies are very different from humans, they are much less different from other jointed-legged animals, such as shrimp. Work on *Hox* genes in shrimp and other crustaceans has begun to reveal how changes in regulation have had an impact on evolution. To that extent, Cuvier was right to confine comparison within strictly defined *embranchements*.

Flies and shrimp are both arthropods, animals whose armoured bodies are arranged in a clearly defined series of segments. Arthropods tend to be classified according to how their segments are disposed. Insects, for example, are very neat – the head region contains a number of mouth parts, and the three segments of the thorax each contain a pair of legs. The second – and sometimes the third – thoracic segments bear a pair of wings. None of the several segments in the abdomen bear limbs or any other kind of appendage. Insect limbs are never branched, and tend to be very specialized. An appendage used for feeding will not be used as a walking leg, or vice versa. Crustaceans, though, have many more, and more diverse, appendages than do insects, and their appendages can be found all along the body, from head to tail. They are often branched and multifunctional, with different parts of the same appendages used for walking, swimming, eating or a combination of all three. In addition, some appendages carry gills for breathing.

Despite these obvious differences, insects and crustaceans are closely related. Some scientists would go further and say that insects evolved from a particular crustacean stock. In other words, insects are a group of crustaceans that specialized in life on land, in the same way that land vertebrates (amphibians, reptiles, mammals and birds) are land-living fishes. This implies that the concise body plan of insects evolved by a process of simplification of the more extravagant crustacean form. Crustacean gills would be useless for breathing on land, and insects have evolved a new system of air-tubes, unconnected with their limbs. Some scientists have speculated that insects transformed the gills of their crustacean ancestor into wings – and that the first insects to fly had many more than one or two pairs of wings. Such engaging ideas aside, it is abundantly clear that crustaceans have many legs on the abdomen – whereas insects have none in this region.

The reason for this difference lies entirely with the evolution of gene regulation. The presence of appendages on the abdomen is determined by a regulatory mechanism involving a *Hox* gene called *Ultrabithorax*. In crustaceans, this mechanism ensures that appendages appear on the abdomen in their appropriate positions. In insects, however, the same mechanism is tuned such that the formation of appendages is suppressed. Insects and crustaceans are sufficiently similar for us to be sure that they both have *Ultrabithorax* genes, and that these genes will be regulated in networks which bear far closer comparison than would be the case with, say, the *deformed* gene in flies and mice.[8] It is thus possible to see how the different forms of flies and shrimp have been shaped by evolutionary changes in gene regulation.

For all these insights – the connection between homeosis and *Hox* genes, the organization of *Hox* genes into clusters, and the implications of this clustering for our understanding of the origin of form – we owe an enormous debt to the dogged persistence of Edward Lewis, finally rewarded in 1995 with a Nobel prize. He shared the award with two other scientists who delved even further into the hidden realm of genetic regulation – deeper even than *Hox* genes.

Hox genes are remarkable morphological generators, but they are not the only game in town: they comprise just one relatively superficial level in a multi-tier system of genetic regulation.

183

These further, occult substrata of regulation were explored in a series of elegant experiments on *Drosophila* embryology, published in 1980 by the German scientist Christiane Nüsslein-Volhard and the American-born Eric Wieschaus, then both at the European Molecular Biology Laboratory in Heidelberg, Germany. Their mission was clear: to make a complete catalogue of monsters which they would use to systematize our understanding of how a fly develops from an egg.[9] As such, their work would fulfil the programme set out by Bateson in *Materials* almost a century earlier.

In a fruit-fly breeding programme of epic scale, Nüsslein-Volhard and Wieschaus set out to discover which of the approximately 20,000 genes in the *Drosophila* genome were involved in development.[10] They discovered a select group of mutations which affected fruit-fly morphology – or rather, the form of fruit-fly maggots, given that most of the mutations were so severe that development beyond a very early stage would have been impossible.

It was evident that *Hox* genes acted in embryonic development at a relatively late stage. Before you can create a homeotic mutant in which legs are traded for antennae, it is first necessary to be able to identify legs and antennae as separate and distinct structures, within the context of a body plan which already exists. *Hox* genes are important, but in the great scheme of things they are the interior decorators that move in to paint an otherwise finished house.

Some mutations have far greater effects, but they are not readily apparent in breeding experiments: like the headless mice, the afflicted embryos are so monstrous that they do not develop very far before dying. For example, there is a mutation in flies called *bicaudal* that creates an embryo with two rear ends stuck together, with no head or thorax. This kind of mutation, resulting in global disruption rather than tweaks in this segment or that, represents a seismic disturbance in the deepest layer of regulation. Nüsslein-Volhard and Wieschaus studied a range of such mutations – grubs with whole swathes of adjacent segments missing; or with only odd-numbered or even-numbered segments missing; or with all the segments present, but the structures within them arranged back to front. By examining and classifying these monsters in a manner befitting Bacon or Paré, the scientists could, at last, piece together the story of how a fly develops from egg to embryo, in terms of genetic regulation.

The most profound level of genetic regulation – and the one that takes place at the earliest stages – relies not on the embryo's own genes but on those of its mother. Proteins produced by the so-called maternal-effect genes map out a global orientation for the fertilized egg, laying down which end will be the head end, and which the rear. It is at this level that mutations such as *bicaudal* act. Once the overall plan is sketched out, a set of the embryo's own genes start dividing the embryo into large but well-defined regions. Mutations in these gap genes result in the absence of several adjacent segments – the entire thorax and most of the abdomen, say, for mutations in the gene *Krüppel*. Only when the body has been divided into these regions can the individual segments be created. The regulation of this seg-mentation process depends strongly on the activities of the pair-rule genes that determine alternating segments (*paired* is one of the genes in this class). Mutations in a pair-rule gene called *even-skipped* result in an embryo that lacks even-numbered segments. For *odd-skipped*, the converse is true. The fourth layer of regulation, above maternal-effect, gap and pair-rule genes – is ruled by the segment-polarity genes, which determine the positions of structures in individual segments. Mutations in these genes are more subtle than the others, and can usually be dis-covered only by careful study of superficial anatomical features.

Only then, once the body is divided into a back and a front, into large regions, and into segments which are in the correct orientation, can *Hox* genes have a stage on which to act. Because they play their parts when much of the important action has already taken place, animals with mutations in *Hox* genes often have a chance of surviving until adulthood – allowing them, in some cases, to survive in the wild so that they can come to the attention of naturalists such as Bateson.

In their revelation of a previously hidden wealth of developmental mutations, Nüsslein-Volhard and Wieschaus presented a show of grotesques that no medieval bestiary could possibly have matched. More seriously, they completed the systematic investigation of the causes of monstrosity started so long before by Paré and continued by Bateson. At a more human level, their results have dramatic implications for human reproduction, and set one aspect of human misery in context. It is thought that a large percentage of early miscarriages – around 40 per cent – may be caused by mutations that affect the fundamental organization of the body. Indeed, mutations that produce gross abnormalities of form will cause spontaneous abortion very early on in pregnancy, often before a woman realizes that she might be pregnant. The mutations studied by Nüsslein-Volhard and Wieschaus were very much of this kind: mutations that killed a fly long before it grew out of the grub stage.

It would be wrong to put too great an emphasis on a picture of genetic regulation as a series of layers, each self-sufficient and discrete but for commands fed to a layer above, with the uppermost layer – the one that is easily visible, and what we would regard as the source of normal variation – constituting the most trivial, at least in respect of the origin of form. Some apparently superficial variation provides a window on to deeper processes. One case involves a gene called *bric-a-brac* that specifies how the pigmentation of the fruit-fly abdomen differs between male and female flies – as superficial a function as might be imagined. But *bric-a-brac* is part of a regulatory network which includes *Hox* genes, and may even have evolved from a homeotic gene itself. On the other hand, the activities of *bric-a-brac* today regulate how flies choose their mates, thus making a connection between the present-day courtship behaviour of flies and genes that control the origin of fruit-fly form.[11]

In another case, the protein products of genes called *engrailed* and *wingless* play an important part in ensuring that the segments of the developing grub are properly oriented. But each of these genes plays other roles at different times during the developmental process. As its name suggests, *wingless* has to do with the specification of wings; mild mutations in the gene result in wingless flies. The name *engrailed* is more obscure. The term comes from heraldry and first appears in this context in the *Morte D'Arthur*, the canonical retelling of the King Arthur legends, in the early fifteenth century. It refers to the condition of edges or borders ornamented with semicircular indentations – a reference to the appearance of segmental boundaries in mutant flies. But in addition to its role in segmentation, *engrailed* plays a part in the development of nerve cells in most, if not all, animals at various stages of their development. Indeed, many scientists believe that the role of *engrailed* in the formation of nervous systems is its primary, even primordial function, and only later was it co-opted into the network of genes specifying segmentation. In this way, genes involved in one regulatory layer may become involved in others, at a different time in individual development, creating a rich network of connections which blurs the edges between individual layers in the hierarchy.

The efforts of Lewis, Nüsslein-Volhard, Wieschaus, and many others have given us a rough understanding of the layout of genes in a network, creating sketch-maps of the origin of form. The next step will be to make those maps come alive:

to investigate how such networks behave as discrete entities in their own right rather than as sums of their genetic components. This is important because networks may have properties all their own which could never be guessed by looking at lists of genes – properties which might shed light on many things that have remained obscure. They may help us to see why it is that monstrosities can distort some parts of the body beyond recognition without affecting other parts; whether the differences between everyday variation and monstrosity are just differences in degree, or reflect a qualitative shift; why variation occurs at all; and how genetic networks can create babies with such startling fidelity and reliability – questions which have perplexed thinkers for millennia.

Here the story of the genome turns from history and becomes reportage. The application of information theory to genetics is a relatively new but rapidly expanding discipline. In the past few years, terms such as 'bioinformatics', 'systems biology' and 'computational biology' have entered the scientific lexicon.

One team of computational biologists, Garrett M. Odell and his colleagues at the University of Washington in Seattle, has been using the network approach in an investigation of *engrailed*, *wingless* and other segment-polarity genes. Taken together, these genes form a small network that ensures that structures in the developing segments of fruit-fly grubs appear in the correct front-to-back order. This 'segment-polarity module' is a semi-autonomous subroutine of the greater network that comprises the entire genome. By studying the module as an integrated entity, rather than focusing on any particular gene or interaction, the researchers hope to learn how genetic networks (and, by extension, whole genomes) behave in general. It is surely too early to put such cutting-edge research in its proper historical context. But what makes the Seattle group's otherwise terse and technical report so engaging is that they clearly knew they were on the verge of important discoveries – knowledge that fuelled a determination to continue despite formidable obstacles along the way.[12]

The scientists' aim was to create a mathematical model of the segment-polarity module, run the program, and see if the computer equivalent of neat, stable segmentation would emerge, in imitation of what actually happens when genes such as *engrailed* and *wingless* interact to order the segments of a fly. That way, they could test the reaction of the network to various disturbances and gain an insight into how networks behave without having to embark on tens of thousands of breeding experiments – potentially collapsing decades of work into a matter of weeks or months. This may sound simple, but the quest to make a functioning virtual segment-polarity network soon ran into serious difficulties.

The first task was to collect everything that had ever been discovered about the interaction between the genes in the segment-polarity module, and assemble these facts into a kind of genetic circuit diagram. In the fly, genes in the segmentation module have rather quaint names. Apart from *engrailed* and *wingless*, there is *hedgehog, patched* and *cubitus interruptus* – names derived from the appearances of flies with mild mutations in the genes concerned. But making a network of genetic interactions requires more than drawing arrows between words such as *hedgehog* and *patched*. More information is needed, specifically about the *strengths* of various interactions, such as how tightly a regulatory protein binds to an operator's DNA, under what conditions and for how long. But our knowledge of such things is very patchy indeed: because regulatory proteins are found in much smaller quantities than are the structural proteins whose abundance they regulate, it is extremely difficult to measure their properties accurately. In any case, most biologists are more interested in the mere fact of the interaction between gene and regulator than in the minutiae of how such interactions are governed at a biochemical level. In the absence of real, known values for these necessary quantities, the researchers labelled each value as an unknown 'free parameter' — an x in the equation. The first essay at a network had no fewer than fifty such unknowns, reflecting our ignorance of the fine details of a system even as well known as the fly segment-polarity network, and making the creation of even a toy version of any genetic network a remote prospect indeed.

At this point, Odell's team hit on a way to answer the question, but from a different angle. Given that we do not know the *actual* values of the unknowns, they asked, might it nevertheless be possible to devise some *plausible* set of values for these unknowns that would make the virtual segment-polarity model work? The answer was a decisive 'no'. The researchers generated thousands of sets of values for the unknowns, fed them into their model and ran the program. The results were not encouraging. In most of the runs, the virtual network never settled down to a state that was stable for long enough to allow segments to form. For those few sets of parameters that produced a stable system, some of the virtual genes were switched on all the time, while others were continually repressed — reminiscent of the mutations in the *lac* repressor or operator of *E. coli* that led to the continual synthesis of beta-galactosidase. In only one run in 3,000 did anything like stable segmentation appear, and only if parameters were tuned very precisely. If just one of the parameters varied by a small amount from its optimum value, the whole system would collapse.

Struck down not once but twice, the scientists now had what proved to be a remarkable insight. They knew that in real life a segment-polarity module must exist, and that it is stable and robust enough to work beautifully in an enormous variety of otherwise quite different creatures. In the language of computer science, a segment-polarity model exists that produces stable solutions when offered an extremely wide range of parameter values. If this were not so, we should not see the results of its work all around us. As Goethe, Geoffroy and Bateson had realized, bodies are often made of repeated segments, whether manifested in the segments of insects or worms, the vertebrae of backbones or even the arrangement of leaves on stems – all suggestive of an underlying natural architectural preference for modular construction. Scientists later confirmed and emphasized this universality, by showing that the same segment-polarity genes that orient the segments in insects organize the development, in discrete compartments, of the brain of human embryos. Segmentation was clearly out there in nature, a production of a network of interacting genes, and was therefore something that could in principle be modelled on a computer. The problem lay not with the approach as a whole, but in the particular model the scientists were using, which was not capable of producing the stable, robust solutions seen in nature. Something was wrong with the model.

It turned out that just two connections had to be added to the circuit diagram of the network to make it function properly. In one, *cubitus interruptus* had to repress *engrailed*. In the other, *wingless* had to activate itself. Crucially, the researchers knew of no biochemical evidence to substantiate these links. All they knew was that these links had to be present for the model to work as a whole. At this point, the researchers crossed an important philosophical divide: from biology as a description of nature to biology as a body of predictive theory. Rather than simply collecting data on how molecules interact – which is really a kind of natural history, albeit on a molecular scale – they were able to use their model as a way of predicting discovery, even directing it. This way of doing science transcends the particulars

191

of biology to become a form of physics, like the particle physics of the mid-twentieth century in which theorists predicted the existence of hosts of the exotic particles they needed to make their picture of the Universe hold together – and experimenters went on to find that the theorists had been right. It is its predictive quality that makes the Seattle group's work a milestone. Happily, the researchers found – after the event – reports describing evidence for these two missing connections.

Once Odell's team added the two links to their model, success was instant and complete. Stable patterns of segmentation seemed to burst from the computer, almost irrespective of the values chosen for the unknowns (reduced from fifty to forty-eight in the revised model). From 240,000 trials, 1,192 – around one in 200 – produced a stable, realistic segmentation pattern, even though each parameter could take any of a huge range of values. At first sight, a success rate of one in 200 seems rather poor, but this should be seen in the context of the huge variability of the parameter values. The researchers calculated that if the model allowed the values of each of the unknowns to vary over just 10 per cent of a hundred- or thousand-fold range, a random search would come up with not one solution in 200 trials, but one in 1,000,000,000,000,000,000,000,000,000,000,000,000, 000,000,000,000. The implication is that parameters in the network can vary by *very much more* than 10 per cent in value, and the network will still produce stable segmentation patterns.

In engineering terms, even a 10 per cent variation represents unbelievable sloppiness. Imagine a factory turning out toasters, each one made from fifty parts, and that the quality control department allows for a 10 per cent error on every part. If such a production line operated for a time equal to the age of the Universe, no functional toaster would ever roll off the end. But put the segment-polarity network in charge, and a perfectly good toaster would made in every few hundred attempts, given the same or even a greater error rate. Clearly, in the real life of genes and proteins, there is no such thing as a 'best' set of parameters: the segment-polarity network is constructed in such a way that an enormous range of values in the unknowns will produce a stable pattern of segmentation.

Odell's group looked at one of the unknowns, the rate at which the protein product of the *wingless* gene diffuses through a cell. Nobody knows the real value of this rate, but that turned out not to matter very much: the model yielded stable solutions when the protein diffused rapidly across cells, and stable solutions when there was hardly any diffusion at all. The model as a whole proved itself to be thoroughly robust, provided the network was wired up correctly. This aspect is the key. The precise base sequence of genes is of little consequence as long as the gene works after some fashion, because the network tolerates huge variations in the efficiency of individual components.

The robustness that Odell and colleagues found in their model network reflects a reality of development that has hitherto remained unexplained in anything more than vague terms. A robust network is defined as one that produces workable results given a wide range of input values. It can tolerate sudden change, and can absorb great damage, yet still produce a viable organism. A good example of a robust network is the Internet, originally designed by the military as a distributed system of computers that would be resistant to large-scale attack. Damage to any one of the Internet's nodes is quickly countered by routing information through other nodes. The system as a whole is therefore able to sustain surprisingly large insults without catastrophic loss of function. Crucially, the function of the Internet is a matter of connectedness, and is not limited by the specifications of any component part – all you need is a modem, and you can connect your basic home computer to a network shared by the most powerful machines ever invented.[13]

Genomic networks, like the Internet, are robust in that they can tolerate enormous flexibility in the efficiency of their various components. They also resemble the Internet in that they are distributed – there is no single, vulnerable hub through which all information must pass. As we have seen, there are subroutines, or modules, in which a set of genes is clearly more connected to others in the same subset than with genes more generally. This makes networks tractable for study, because it is possible to examine modules more or less in isolation – the Seattle group's work on the segment-polarity network is a case in point. But modularity is a desirable property of robust networks, because damage to any one part of the network can be contained relatively easily, without compromising the rest of the network.

Modularity explains why it is possible to produce dramatic monstrosities in which one part of an animal is distorted, duplicated, substituted for another part – or even completely absent – while other parts develop quite normally. *Lim1*-mutant mice have no heads, but their limbs, tail and bodies are normal, at least from a superficial inspection.[14] In the classic *antennapedia* mutation as studied by Bateson in insects caught in the wild, the monstrosity does not spread beyond the leg-like antennae, and the affected insect is otherwise quite normal. This modular damage-limitation strategy is evident even with the more drastic mutations described by Nüsslein-Volhard and Wieschaus. A mutation in a pair-rule gene such as *even-skipped* may cause the deletion of even-numbered segments, but the odd-numbered segments form regardless. Likewise, a mutation in a gap gene such as *Krüppel* may result in the deletion of large regions of the body, but segments still form in the parts that remain. That is, mutations in such genes do not result in the total collapse of the network, creating embryos which are, invariably, formless blobs of protoplasm. The modular construction of networks ensures that the unaffected parts of the body develop normally, as if in complete isolation from the total devastation just a few cells' breadth away.

The toleration of networks for enormous ranges in the efficiency of their individual components could also help explain the nature and existence of variation. Darwin was the first to understand the crucial importance of variation to evolutionary change. As such, he was keenly aware that his theory of natural selection was utterly dependent on the existence of variation, and that natural selection could not function without some mechanism to replace the variation culled in each generation by selection's scythe. Darwin's failure to advance such a mechanism contributed to the eclipse of natural selection at the end of the nineteenth century, creating a void that could be filled only with the kind of vacuous evolutionary storytelling that drove Bateson and Morgan towards a more experimental approach to the problem. Nobody can doubt the importance of what happened next: the subsequent advances of genetics, the chromosome theory, the rehabilitation of Darwinism in the 'modern synthesis', the discovery of the structure of DNA, and so on were a direct result of the dissatisfaction so eloquently expressed by Bateson in *Materials for the Study of Variation*. As a result, we are now able to study variation in all its forms, from individuals in populations down to the interactions of genes and proteins.

Given the (rightful) domination of evolution by natural selection as the pre-eminent theory of why the natural world is the way it is, it is all too easy to explain the existence of variation only in terms of its contribution to evolution. Mutations, whether produced by X-rays or just by accident, generate variation, which is then acted on by selection. Sex, based on the same phenomenon of recombination that was crucial in validating the chromosomal theory of heredity, generates far more variation than is possible with random mutations alone, providing natural selection with an even broader palette of variation with which to work. The evolution of sex a couple of billion years ago, in some simple unicellular creature, stoked the fires of evolution and opened the way to the appearance of complex, multicellular life.

However, to say that variation is necessary for evolution, and that more variation made evolution faster, is in itself no argument for the prior existence of variation or an explanation for its nature. To view variation simply as the handmaiden of evolution is to leave a number of questions unanswered. What explains the range of what we think of as 'normal' variation? Why isn't the range of normal variation greater than it is, or less? Indeed, why do creatures vary at all? Why do monsters seem to violate our innate sense of what constitutes the normal range of variation? Does this sense have any basis in fact? The emerging network view of genetic regulation allows me to propose – albeit tentatively – a reason why variation exists, and why it is particularly important for complex, multicellular creatures such as ourselves.

Monstrosities such as flies with extra wings and mice without heads might be explained in terms of disturbances to the connectedness – the topology, if you like – of genetic networks that are otherwise robust and modular. But variation – that is, what we think of as the normal range of variation expected in otherwise fully functional, healthy organisms – might also be regarded as a consequence of networks that allow a very large degree of latitude in the efficiency of their individual components. As the Seattle group showed, a network can take components of almost any quality or ability and channel their interactions to produce more or less the same product, provided the network is wired up in the correct way. Given the latitude allowed for the input values, it is to be expected that the output of a robust network — in this case, living organisms — might also be varied, although to a lesser degree, and at least in respect of characteristics that would not harm the overall function of the network — in other words, characteristics that correspond to normal variation.

Variation — the kind of variation without which natural selection cannot work — might, therefore, be simply a by-product of the particular way in which the genome directs the formation of each new organism — that is, as a robust network of interacting genes, each of which can vary hugely in its own properties and propensities. But this begs an even deeper question: why is it that the genome is organized in just this way, to produce organisms that are variously flawed, when it could have created organisms that are all identical, and all perfect — the nature-philosophical ideal, rather than the messy reality? The reason, I think, illustrates the pervasive power of natural selection. What is at issue here is not so much the *degree* of variation we see in natural populations, but the very *existence* of that variation — and this existence is itself subject to natural selection.

Genomes first started to create multicellular creatures more than a billion years ago. Until that time, the world was populated by unicellular creatures. When a single-celled creature divides into two, it produces two complete, adult creatures more or less instantly, with no need to create complex form from a formless egg. Unicellular creatures, therefore, do not really have 'development'.[15] The evolution of multicellular creatures, therefore, raised a novel requirement: the need to direct matters to ensure the safe passage between egg and adult; and to do this efficiently, and with the best chance of achieving a viable result, rather than a monster or nothing at all.

In evolution, all that matters is that an organism develops from a fertilized egg until it reaches adulthood, so that it can then reproduce and pass on its genomic heritage to the next generation. However, in all but the very simplest multicellular creatures, the course of development is so intricate that a single slip at any stage has the potential to threaten the integrity and viability of the whole. Given the uncertainties of the real world, any organism that develops according to an overly strict, step-by-step procedure has very little chance of reaching term, let alone sexual maturity. For every single, perfect organism produced in this way, large numbers of embryos might have started on their uncertain courses, just so that one might emerge as an adult without suffering the fatal consequences of some mishap along the way – a mishap which, given the complexity of the overall process, and the dependence of each step on the one preceding, would be virtually guaranteed to happen. Against potentially terrible odds, any organism whose genome evolved a strategy of development that improved this dismal success rate, even by a tiny amount, would inherit the Earth.

That is indeed what happened – the future belonged to those creatures whose genomes were organized flexibly, so that their development could not be completely derailed by relatively minor disasters. These organisms would have been more successful at reproduction than those with less flexible genomes, and so would have been favoured by natural selection in Darwin's canonical struggle for life. With the passing aeons, selection would have continued to favour the flexible, so that the most successful genomes were those whose activities were organized into decentralized, modular networks, able to tolerate an ever greater range of variation in the properties of their components – at the small cost of some residual variation between the individuals so produced. Identikit perfection, if it were ever achieved, no longer exists.

More than a billion years later, the consequences of this primordial selection for flexible, network-based genomes are all around us, in the birth of yet more human beings with each passing minute, each one minutely formed, created with a reliability and flexibility which almost defies belief. As a consequence of this very reliability and flexibility, each new baby is a unique individual with its own potential to be an athlete, an architect or an actuary; its own predisposition to disease and response to medicine; its own likes, dislikes, needs, hopes, wants and dreams.

I suggest that variation, for all that we take it for granted, and for all its subsequent utility in evolution, is a by-product of the truly amazing piece of network engineering that creates form from the formless. And yet, the existence of this variation might be the price that each of us pays for having been born at all.

11

Scars of evolution

The view of the genome as a network makes premature any claim that the secret of humanity might lie in the sequence of DNA that makes up the human genome. This is a consequence of the distributed nature of networks: at no point in the human genome can you find clear instructions to make a baby. At the most basic level, one sequence of DNA looks very much like another. Even when comparing whole genes, rather than sequences of DNA, it is often extremely hard to say that a particular *Hox* gene, say, comes from a man or a mouse. This is true for the comparison of any individual genes between different organisms, even if the organisms look much more different than such comparisons might suggest. Expressed in this way, the network view validates the instincts of the ancients that the genome is important for what it does, rather than for what it is made of.

Now that we are confronted with the physical reality of the genome, however, it is pertinent to ask whether the sequence of the human genome can shed any light at all on what it means to be human. In a way, it can. Not, of course, in that the secrets of human dreams and desires might be decoded from a raw sequence of bases, but because the genome is constructed in such a way that it tells its own story, showing how each human being, moulded from an egg into a recognizable human shape in just twenty-eight days, is a product of billions of years of evolution.

If we look into the human genome for the first time, expecting it to be an orderly arrangement of genes, we will be in for the same kind of shock as awaits a tidy householder who returns home to find it ransacked. The human genome consists of some 3 billion bases, but less than 5 per cent of this enormous total actually consists of genes that code for anything. The genes are all but buried in so-called junk DNA – apparently meaningless stretches of highly repetitive sequences known as satellite DNA, defunct genes, the wrecks of old viruses and other genomic detritus. Any given stretch of a million bases will contain only eleven genes, but this is an *average* figure. A few parts of the genome are crowded with genes, but some regions appear to contain virtually no genes at all. Estimates of the number of human genes vary between 31,000 and 39,000. The estimate is this vague, even given the relative completeness of our knowledge of the human genome, because human genes, even where present, are broken up into many small pieces and are very hard to spot in a mass of DNA sequence.

About half of the human genome consists of transposable elements – parasitic stretches of DNA derived from viruses which, at face value, do little other than encode their own existence. The bestiary of junk DNA contains several varieties of transposable elements, but of particular interest are the so-called long interspersed elements, or LINEs. There are around 850,000 LINEs in the human genome, comprising around a fifth of its DNA.

The origin of LINEs is to be found among viruses, in particular a class called retroviruses. Unlike most organisms, whose genomes are made of DNA, the genomes of retroviruses are made of RNA. Retroviruses use an enzyme called reverse transcriptase to make DNA copies of their genomes, which they then insert into the DNA genome of their hosts. Perhaps the most notorious retrovirus is the human immunodeficiency virus HIV-1, which causes acquired immune deficiency syndrome (AIDS). Although LINEs are made of DNA, they spread around the genome by copying themselves into messenger RNA, and then copying this back into DNA in a new location in the genome. Retroviruses, like all viruses, have genes that encode various proteins – LINEs have reduced this to a complement of just two: one that encodes the reverse transcriptase enzyme, the other an enzyme called endonuclease, which makes a cut in the host DNA to allow the copied DNA of the LINE to sew itself into the host genome.

LINEs are often accompanied by members of another class of junk DNA, the so-called short interspersed elements, or SINEs, which can spread themselves through the genome only by parasitizing LINEs. SINEs are much shorter than LINEs – between 100 and 400 bases, compared with around 6,000 for a LINE – and have no genes at all. All they contain is a stretch of DNA designed to gain the attention of the host's transcription enzymes. In short, SINEs are nothing more than notices saying 'Me! Me! Me!' Once transcribed into lengths of messenger RNA, SINEs lie in wait for the endonuclease of a LINE to cut a section of host DNA ready for the insertion of a LINE, and for LINE reverse transcriptase to convert them into DNA. As a result, a LINE is usually bracketed by one or more SINEs, in the same way that a cat might be accompanied by a retinue of fleas. SINEs do not derive from viruses, but from certain kinds of defunct gene. The most important SINE in the human genome is called *Alu*, and over the course of millions of years copies of *Alu* have spread through the genome so that this element is perhaps the most important genomic parasite in human evolution. Of 1.5 million SINEs in the human genome, 1.3 million are copies of *Alu*, accounting for almost 11 per cent of the entire genome.

LINEs and SINEs go back a long way together. A particular species of LINE, called LINE1, has existed for around 150 million years: it has been a part of the human genome since before we were human – indeed, from when we were primitive mammals. *Alu* is parasitic on LINE1, and has been a regular feature of the human genome for 80 million years. Even before that, the most important and active LINE in the genome was LINE2, parasitized by a SINE called MIR. But LINEs have a finite lifespan, albeit measured in tens of millions of years. When all available copies of LINE2 became mutated or otherwise disabled, it effectively became extinct. And when that happened, MIR was also doomed to extinction. The days of LINE1 and *Alu* may also be passing. LINE1 is currently the only active LINE in the human genome, but the definition of activity is somewhat relaxed: it appears to have been dormant since before the remote ancestors of humans climbed down from the trees. *Alu* went through a burst of activity around 40 million years ago, but has done very little since.

LINEs and SINEs do nothing except promote their own existence, and it would therefore be very easy to dismiss them as genomic freeloaders. However, they are as much a physical part of the human genome as are bona fide genes. The dichotomy between genes and junk, and our sense of shock that so little of the genome actually consists of genes, derives from the misconception that the genome is no more than a catalogue of parts from which we ought to be able to learn, as if by direct apprehension of the sequence, what it means to be human. Such a view naturally dismisses junk DNA for what it seems to be. But when the genome is considered as a whole, the junk DNA is seen to provide, by its very preponderance, the contextual landscape, the scenery against which the genes play their parts.

Kevin Padian of the University of California, Berkeley – palaeontologist, historian of science and Thomas Hardy scholar – has discussed how the landscape of Hardy's Wessex is almost a character in its own right, whose scale, formed over millions of years, dwarfs the parochial and short-term concerns of the characters who live within it, and whose natures and fates are, to an extent, determined by its topography.[1] Much the same could be said of the relationship of genes to the genome they inhabit. The vast stretches of simple, repetitive DNA sequences form an immense and centuried landscape that frames the small and parochial dramas of the genes within it. If the junk really were detrimental to the activity of genes, then evolution would have contrived a way for us to shed it: it is the junk as much as the genes that makes us human, in the same way that the characters of Tess Durbeyfield and Angel Clare are shaped and given meaning only in the context of the landscape of Wessex.

Some perspective may be offered in the unlikely guise of the genome of the puffer fish (*Fugu rubripes*), otherwise known as an ingredient in Japanese cuisine potentially so toxic that chefs must be specially trained in its preparation.[2] The *Fugu* genome contains the same number of genes as found in humans (about 31,000), but in a space just one-eighth the size, being relatively free from junk. It might be far-fetched to say that without junk DNA we'd all be sushi, but the comparison between people and puffer fish emphasizes the point that it is this apparently redundant DNA that acts as the landscape in which the genes must operate, a topography so distinctive that it modifies the entire character of the genome and its ultimate creation, the organism.

The landscape of the human genome, as of Wessex, is a

varied one. Some parts of the genome consist almost entirely of junk. Ninety-eight per cent of a 200,000-base segment on chromosome X consists of repetitive DNA – LINEs, SINEs, and so on, in great drifts. Within this there is a section of 100,000 bases of which 89,000 consist solely of LINE1 elements, lined up end to end. Elsewhere, the genomic explorer will run into desert dune-fields of *Alu*, or stagnant marshes of MIR surrounding lifeless lagoons of LINE2. Other parts of the genome are relatively free from such detritus, like mountain-tops poking their bald heads above the jungle. Repetitive sequences almost fade out altogether in the four *Hox* clusters. Being closely regulated, these genes depend for their function on being able to interact with one another without hindrance, so the DNA in these regions is almost entirely free of clutter.

In its physical structure, the genome offers no clue to its function. It is hard to see from its structure how it could direct the formation and maintenance of any particular organism, let alone a human being. As to its history, the forces that shaped the genome are akin to those that create landscapes – time, circumstance, and chance. The evolution of the human genome had no clear end in view: there was no force directing evolution towards the human state. Throughout biological history, the activities of genes, and networks of genes, have been modulated by an accumulation of transposable elements and other non-genic DNA, and the balance between these various genetic constituencies has been set by the needs of the organism at any given time. There is a reason why human beings have so much junk DNA, just as there is a reason why puffer fishes have so little. In each case, the reason lies in the efficient function of the organism as either human or puffer fish.

The same point can be made, over and over again, for the billions of years of the evolution of the genome. In the course of human ancestry, each organism in which the genome found itself at any given moment, whether bacterium in the primeval depths, or primate in the branches of great forest trees, had to adapt to its own here-and-now, and the genome adjusted itself to suit. So, although the genome shows a trend towards increasing bulk and complexity over the past few billion years, this has been subject to variation – and some spectacular reverses. Despite the impression of disorder and waste conveyed by a genome in which 95 per cent is junk, organisms maintain only as much genome as they need. This in itself explains why junk DNA forms a necessary part of genomes when considered as whole things; it also alerts us to the possibility, indeed the fact, that many key episodes in the evolution of the genome were accompanied by wholesale losses of genetic material as well as gains. The evolution of LINEs from retroviruses, and of SINEs from functioning genes, shows how some organisms, particularly parasitic ones, shed genes as they adapt to new modes of life, and this phenomenon is seen again and again, at every level. Time has collected the myriad accidents and quick fixes that allowed the various ancestors of humanity to survive at any given moment, archiving them in a genome which is at once the creator of a new organism and a testament to its evolutionary history. In this sense, if in no other (and particularly not as regards any sense of manifest destiny), the genome is very much the microcosm that measures the macrocosm, as prescribed by the nature-philosophers. What follows is a brief account of the history of the genome, showing how the scars of evolution are still borne by organisms alive today – humanity not the least – scars which stand as monuments as well as being parts of functioning genomes which continue to create and maintain organisms of all kinds.

The Earth is believed to have formed approximately 4.5 billion years ago. Life was quick to appear: the first (debatable) traces of living things have been dated to approximately 3.8

billion years ago.[3] These are no more than quirks in the chemical nature of the oldest sedimentary rocks, and the precise form of the earliest living things is therefore unknown. The simplest self-reproducing entities that now exist are the viruses, which have no life until they infect a cell. Because of their simplicity, viruses provided the first genomes to be sequenced in their entirety. Unpicking the genomes of viruses, one base at a time, allowed researchers to develop the technology that was, ultimately, to read the human genome itself. The first genome to be sequenced, in 1977, was that of a virus of bacteria known by its rather gnomic number, ψX174. This virus has a DNA genome 5,375 bases in length, containing just nine genes. The genome of Simian Virus 40 (SV40), which as its name suggests is found in monkeys, was published in 1978, and bacteriophage lambda – the partner of *E. coli* – followed in 1982. The genomes of more than 600 viruses are now known in their entirety.[4]

The existence of RNA genomes, in retroviruses, suggests that RNA might have preceded DNA as the carrier of genetic information. In almost all organisms apart from retroviruses, genetic information resides in DNA, but this information cannot be accessed without enzymes, which are made of proteins. Yet the instructions to make proteins are contained in DNA, which cannot be read without proteins – and so the enzymes come and go, towards *absurdam, reductio*. RNA, being the intermediary between DNA and proteins, may offer a solution to this paradox. We know that RNA, like DNA, carries genetic information. However, there are many kinds of RNA molecule that fulfil other roles – molecules such as transfer RNA. The existence of these and other kinds of non-coding RNA demonstrates that RNA is in many ways a more versatile substance than DNA.

In some organisms, sections of messenger RNA which are not destined to be translated into sequences of amino-acids can remove themselves from the messenger RNA sequence, by acting as their own removal enzymes. Such self-splicing RNA molecules have some of the properties of enzymes, as well as containing genetic information. This dual role may have been essential to the origin of life, and to the birth of the genome: it could be that the first steps to life were taken by molecules such as RNA which carried genetic information but could also use their own chemical properties to catalyse their own reproduction. Many researchers have explored the implications and consequences of this recursive, by-its-own-bootstraps model of genesis, imagining an 'RNA world' before the days of DNA and proteins.[5] Perhaps the first catalytic RNA-like entities deposited or archived their genetic information in stable, long-lived DNA, while at the same time inventing the genetic code to create proteins, which as catalysts offer far greater versatility than RNA.

Where did the earliest life forms live? Darwin envisaged a 'warm little pond', and many researchers have since built on this view, showing how solutions containing simple chemicals could yield complex organic molecules if energized by lightning or radiation. Others have suggested that life might have begun in the deep sea, in the hot, highly pressurized environments of deep-sea hydrothermal vents. However, there is a problem with the origin of life in such settings. For a molecule to build itself up from smaller building blocks, it must be immersed in a very thick broth of these ingredients for this accretion process to work effectively. In a dilute solution, such as pond water or the sea, a self-assembling, long-chain molecule is more likely to shed newly added components than gather any from the environment.

A more attractive scenario for the origin of life was proposed in the 1970s by the British scientist Graham Cairns-Smith, who suggested that molecules would have a much better chance of meeting if they had some defined meeting place, rather than relying on random collisions in a dilute solution.[6] For example, the surfaces of particles of certain minerals found in clay provide ideal venues for such molecular assignations. Clay particles have rough surfaces, full of crevices in which molecules may find one another in undisturbed proximity, allowing them to react with each other chemically – reactions that might not take place at all were the individual molecules free to float in the environment. Small segments of RNA on clay surfaces could have reacted with one another, building longer chains than would have been possible in dilute solution, and so created the first genome – with the mineral surface acting, in effect, as the first enzyme.

The trick, though, was containing these delicate germs of living material so that they became concentrated and their interactions self-sustaining. Cells function because their membranes ensure that chemicals cannot stray very far, and because they actively expend energy to acquire more ingredients from outside. At some point in the early evolution of life, cell membranes were created. In essence, cell membranes are not so very different from soap bubbles, or scummy layers of fat on water. In certain conditions, fatty substances form bubble-like structures rather easily, and these bubbles can grow and bud off smaller bubbles in a very life-like way. Some theories about the origin of life envisage this growth and budding process becoming linked with the growth and reproduction of molecules containing genetic information. Eventually, then, the first cells appeared.

Among the smallest and simplest of all single-celled organisms alive today are bacteria called mycoplasmas, which cause a number of diseases in humans and animals. Mycoplasmas have genomes commensurate with their minute size. The genome of *Mycoplasma genitalium* is 580,000 bases long, and contains around 470 genes, and is the smallest bacterial genome yet sequenced. This genome is thought to be almost as small as a genome can be and still be capable of supporting an independent cell, and is therefore of great interest to scientists trying to establish the smallest possible number of genes that a cell might require to function. This simplicity must, however, be qualified. Mycoplasmas are parasites that cannot survive outside their hosts, which supply many of their needs. As a consequence, the simplicity of their genomes is not primordial, but highly evolved: these genomes represent the much-reduced, decayed remnants of the genetic complement of the more complex, free-living, ancestors of mycoplasmas. The last truly primordial, primitive cell probably lived billions of years ago, and the size of its genome cannot therefore be known. Even so, the genomes of mycoplasmas provide a kind of notional lower limit on the size of a genome that can support any kind of cell, whether parasitic or free-living.

The genetic material of all bacteria, including mycoplasmas, is invariably made of DNA, usually in a single, circular chromosome. Occasionally, this chromosome is augmented by one or more smaller DNA elements, which are also usually circular, called *plasmids*. Plasmids often contain genes which are important to the life of the bacterium, and bacteria swap plasmids and other segments of DNA with insouciant ease. Because of this flexibility, bacteria have been able to colonize every imaginable habitat, from the upper atmosphere to oil-rich rocks deep within the Earth, from radioactive waste tips to the human gut. Bacteria remain the most successful living things on Earth.

For all their ubiquity and variety, bacterial genomes have many things in common. They are generally very concise, with very little of the junk that is found in human DNA, for example. Typically, less than 1.5 per cent of a bacterial genome consists of such repetitive sequences. The first complete bacterial genome sequence to be published was that of *Haemophilus influenzae*, as recently as 1995. This representative bacterium, which is a cause of meningitis in humans, has a genome 1.8 million bases long, and contains 1,743 genes.[7] Since then, dozens of bacterial genomes have been sequenced, from organisms both commonplace and exotic; from the simple, parasitic mycoplasmas mentioned above to free-living and flexible soil bacteria with formidably complex metabolisms and genomes to match; from the organisms that cause diseases such as cholera and plague to creatures known only from sulphurous industrial slag heaps and hot springs.

Their diversity illustrates that the genomes of organisms are there to serve their current needs, rather than to represent, as if for our benefit, staging posts in the evolution of humans. Genomes do not show a continuous gradation in size, but can vary enormously according to the lifestyle of the organisms in which they reside. The genomes of mycoplasmas may have been reduced as a consequence of parasitism. Because parasites rely on their hosts for many of their needs, they can get by with fewer genes of their own. That evolution is marked by the decay of genes as much as their growth is most dramatically illustrated by a comparison of two bacteria, *Mycobacterium tuberculosis* and *Mycobacterium leprae* – closely related to each other, and responsible respectively for tuberculosis and leprosy, among the most feared diseases ever to have afflicted humanity. The genome of *M. tuberculosis* is around 4.4 million bases and contains about 4,000 genes. The genome of *M. leprae*, on the other hand, is significantly smaller, around 3.3 million bases, with 2,700 genes – at least 1,100 of which are non-functional duds. Compared with the genome of *M. tuberculosis*, that of *M. leprae* seems to have been scarred by extensive rearrangements, leading to irreversible damage to many genes, and their subsequent decay by mutation into unintelligibility and eventual loss.

Because of the decrepitude of its genome, the leprosy bacterium is virtually unable to digest anything, having lost the capacity to make most of the necessary enzymes, and it is barely able to process even those substances that its fragile constitution can cope with. The leprosy bacterium is an invalid living in the domain of horror it creates – but it endures, because its genome, while as rotten as the flesh of its victims, is all it needs to get along.

Evolution is marked by the decay of genes as much as their robust growth. And yet in such genomic decay lies the seeds of the next stage in genomic evolution – the transition from bacteria to the larger and more complex cells of higher organisms, whose genomes are made from the decayed and abbreviated genomes of various bacteria. Take, for example, bacteria of the genus *Buchnera*, which can survive only inside the bodies of aphids, and only then within special cells called bacteriocytes, evolved specially for the purpose. *Buchnera* is a relative of *E. coli* but with a genome one-seventh the size: a drastic case of gene loss. The bacterium can make amino acids which its aphid host is unable to manufacture, but it lacks all genes for defence, or for components of the cell surface. Given that the bacterium is well protected within a cell provided by the host, such externals are in any case quite unnecessary. But this raises another question. *Buchnera* has evolved to provide an essential service to its aphid hosts, in the course of which it has completely sacrificed its independence as an organism. So where, then, does aphid end and bacterium begin? Should the diminutive genome of *Buchnera* be considered a kind of satellite to the aphid's own genome – particularly as the host cannot reproduce without it?

In bacteria, the genetic material floats freely within the cell. In the cells of many other organisms – animals, plants, fungi, algae and a host of single-celled organisms – the genome is confined within a body inside the cell called the nucleus. Organisms with cells like this are called eukaryotes. Besides having their genome confined to the nucleus, the cells of eukaryotes are divided and subdivided into a number of compartments, and contain other bodies, organelles ('little organs'), which contain their own genomes, small yet separate from the nucleus. For example, nearly all our cells contain small sausage-shaped bodies called mitochondria, responsible for the function of respiration – that is, generating energy from food. Mitochondria have their own genomes, much smaller than the genome that resides in the nucleus. So it is with chloroplasts, those bodies in the cells of plants and many algae which can generate energy from sunlight, and whose light-trapping pigment – chlorophyll – gives plants their characteristic green colour.

The eukaryote cell is rather like the bacteriocyte in an aphid, taken a stage further. Indeed, the bacteriocyte gives an insight into how eukaryote cells evolved – as collectives of different bacteria, working together until, over the ages, the bacteria became inseparable and mutually interdependent. Indeed, many bacteria work together today in loose assemblages called biofilms, found in places as diverse as the ocean floor, the plaque on your teeth, and the mucus-clogged lungs of people with cystic fibrosis. In biofilms, however, the bacteria are more or less independent. The eukaryote state, in which the individual bacteria have been subsumed into a greater whole, represents a biofilm taken to the extreme.

The first eukaryote cells appeared sometime before 2 billion years ago, and eukaryotes underwent a huge evolutionary diversification about a billion years ago. This revolution paved the way for today's eukaryotes – algae, fungi, plants and animals. The reason for the eukaryote revolution may have been the consolidation and elaboration of the nucleus, originally just one member of the collective, as the primary information centre of the cell, to which many of the original genes of the members of the collective migrated. The nucleus became the repository for genetic information to such a degree that many members eventually lost their genomes altogether, their activities being devolved to the central nucleus. The modern cell was born, and with that, the modern genome.

The genomes of around 200 mitochondria, chloroplasts and other such bodies have now been sequenced in their entirety, allowing us to make detailed comparisons with the genomes of free-living forms, and allowing us to glimpse the long-vanished bacterial collectives that gave rise to eukaryotes. Chloroplasts, for example, have many resemblances to the cyanobacteria, a group of bacteria that harvest the energy from sunlight. Because of the presence of chlorophyll, tracing chloroplasts to cyano-bacteria is relatively easy.

The origin of mitochondria is harder to fathom than that of chloroplasts. One clue comes from the rickettsias, another group of bacteria in which parasitism has led to extensive genomic decay. A curious feature of one of these organisms – *Rickettsia prowazekii*, the agent of epidemic typhus – is that the bacterium depends entirely on oxygen for its respiration. This may not seem so strange to humans accustomed to breathing oxygen, without which we should soon die, but such a dependency is most unusual in bacteria. Indeed, oxygen is lethally poisonous to many of them. Why, then, should *R. prowazekii* be so differ-ent, so reliant on oxygen? The answer could lie in a deep, shared heritage with mitochondria, for which a commitment to oxygen-based respiration is not only a characteristic feature, but their *raison d'être*. The main function of the mitochondria in our cells is as furnaces, using oxygen to burn our food, thus releasing the energy we need. Could mitochondria be the remnants of rickettsia-like organisms, participants in the bac-terial association that eventually led to the eukaryote cell?

The link is tantalizing, but not entirely convincing. If the mitochondria of eukaryote cells were once rickettsias, evolution has since all but buried the connection under an oppressive weight of change. The remnant genomes of modern mitochondria are very small, and the genome of R. prowazekii, although much reduced, is still ten times the size of the most bacterium-like mitochondrial genome known, the 69,034-base genome of the mitochondrion in a single-celled eukaryote called Reclinomonas americana.[8]

And what of the nucleus, the organelle that came to dominate all others? In the origin of the nucleus lies the birth of the genome as we know it today, yet the ultimate source of the nucleus remains a mystery. The most promising clues come from what must be the most bizarre organisms of the living world. These creatures are called archaea. Superficially, archaea resemble bacteria in that they are minute, single-celled organisms whose genomes float freely in the cell (bacteria and archaea are collectively known as prokaryotes). Archaea are abundant everywhere, although they are often associated with extreme habitats, such as hot springs or deep underground – habitats sometimes thought by some to exemplify the earliest known habitats of life on Earth, hence the name 'archaea' – the ancient ones.

But things are not as they seem. Analysis of archaeal genomes reveals a curious split. Most archaeal genes resemble those in bacteria. However, many others look more like genes seen in eukaryotes – particularly those genes concerned with genomic housekeeping chores, such as transcription and translation. These are the very functions in which the eukaryote nucleus is most closely involved: could it be that the eukaryote nucleus comes from some archaeon that gathered a retinue of bacteria, to their mutual benefit? Over many millions of years, the archaeon came to dominate this loose bacterial collective, sucking genes from its bacterial associates, demoting them to the status of subservient organelles. *Rickettsia*-like organisms became mitochondria, and cyanobacteria were enslaved as chloroplasts.

If so, then eukaryotes such as ourselves can trace the bulk of our genomes directly to the ancient relatives of some of the strangest organisms alive today. One such is *Thermoplasma acidophilum*, which lives inside the hot and strongly acidic environment deep inside large slag heaps by coal mines. Another is *Archaeoglobus fulgidus*, a nuisance in the kind of place that even the most verminous pests might not be expected to survive at all. It lives on sulphur in oil wells deep underground, producing iron sulphide as a waste product, corroding the iron and steel pipework used in the oil extraction process. Organisms like these could have been the distant ancestors of the eukaryote nucleus, and, ultimately, the human genome.

The transition from the prokaryote to the eukaryote state – from a collective of more or less independent organisms to an integrated cell with a nucleus – represents a quantum leap in terms of structure. In general, eukaryote cells are larger, more complicated and more highly organized than bacterial or archaeal cells. However, this complexity is not matched by gene numbers in any straightforward way. The number of genes in archaeal or bacterial genomes varies some sixteenfold, from the 500-gene complement of mycoplasmas to the relatively gigantic 8,000-gene genome of *Mesorhizobium loti*, a soil bacterium which forms elaborate symbioses with the roots of pea plants and other legumes. This genome is larger than those of some eukaryotes, including those of the yeast *Schizosaccharomyces pombe* (4,824 genes) and *Encephalitzoon cuniculi*, a tiny eukaryote parasite which, like rickettsias, lives inside larger cells. This genome contains just 2,000 genes, a number undoubtedly reduced by gene decay and loss. The qualitative difference between bacteria and eukaryotes may be large, and significant in terms of the sweep of evolutionary history. But in the end, the numbers of genes in a genome depends very largely on the lifestyle of the organism served by that genome.

The 13-million-base genome of *S. pombe* is the smallest known for any free-living eukaryote, and therefore gives us a good idea of the minimal gene requirement for a eukaryote. Although the genome of *S. pombe* is no bigger than that of many bacteria in terms of the number of genes it possesses, it contains much more DNA. And although bacterial genomes contain relatively little junk, some 43 per cent of the genes of *S. pombe* are fragmented by spacers of junk DNA known as introns – rare in bacterial genomes. As you might expect, many of the genes of *S. pombe* govern functions peculiar to eukaryotes, such as managing and organizing the various compartments within the cell; creating the internal network of bracing struts, the cytoskeleton (a uniquely eukaryote feature unknown in bacteria); and coordinating cell division so that the chromosomes divide at the same time as the rest of the cell (bacterial cell division is less organized and precise).

After the invention of the eukaryote cell, the next great transition in living forms came with the aggregation of these cells into integrated, multicellular organisms, which happened on many separate occasions between 1,000 and 600 million years ago. Compared with the bulk of life's diversity, multicellular eukaryotes such as animals and plants seem like rare aberrations created by colonial, unicellular eukaryotes. Indeed, green plants are close relatives of unicellular green algae, and with a sufficiently broad perspective, animals look like elaborate colonies of choanoflagellates, another group of single-celled eukaryotes.

The transition from unicellular to multicellular is not as abrupt as that from prokaryote to eukaryote. Indeed, some habitually unicellular eukaryotes may occasionally adopt a multicellular habit. Slime moulds, for example, are simple

217

eukaryotes that occur mostly as single-celled, amoeba-like organisms, crawling around in woodland shade. Occasionally, thousands of these cells aggregate to form a multicellular, slug-like creature, capable of coordinated forward motion. This creature settles down and differentiates, sending into the air a stalk which terminates in a mushroom-like fruiting body. On ripening, this stalk scatters spores, each containing a single cell. The life cycle of the slime mould may seem strange, almost alien – but the fact is that all multicellular eukaryotes are unicellular at some stage in their life cycles. Even human beings go through a brief stage as a unicellular organisms. After all, what are eggs and sperm but single-celled bridges between generations of multicellular organisms?

The transition to the multicellular state was nevertheless accompanied by a marked increase in the number of genes. The first multicellular eukaryote to have its genome completely sequenced was that of the roundworm *Caenorhabditis elegans*.[9] This may seem an obscure choice. However, *C. elegans* has long been popular in laboratory experiments on longevity (as I mentioned in a previous chapter) and other aspects of the genetic control of development. This is partly because its development is peculiarly regimented. Each and every normal adult *C. elegans* has *precisely* 959 cells (excluding sex cells). The lineage of each of these cells can be traced back to the single cell of the fertilized egg. The 97-million-base genome of *C. elegans* contains about 19,100 genes – many more than in yeasts or bacteria. It also contains more junk. Only 27 per cent of the DNA – a little more than a quarter – comprises functioning genes.

To illustrate that there is no simple relationship between the apparent complexity of an organism and the number of genes in its genome, the genome of the fruit fly, *Drosophila melanogaster* has *fewer* genes than *C. elegans* – around 13,600 – even though

it has elaborate organs such as legs, wings, eyes, and so on that
C. elegans patently lacks.[10] Even accepting that gene number is
not simply linked to complexity, it is possible to establish some
very rough, ballpark figures. A genome of at least 1,000 genes
is required to maintain a simple, free-living bacterial cell,
whereas the simplest free-living eukaryotes require around
4,000 genes. The transition to multicellularity was accompanied
by another leap in gene number. If we are to judge from worms
and flies, it requires at least, say, 10,000 genes to make an
animal.

As far as we humans are concerned, the next stage in genomic
evolution was associated with the appearance of vertebrates –
the group of animals which includes ourselves. The earliest
known fossils of vertebrates – very simple, fish-like creatures –
come from rocks in southern China laid down around 530
million years ago,[11] so the actual origin of vertebrates must lie
even further back in time. That vertebrates are inherently more
complicated creatures than fruit flies or worms cannot be
doubted, and as we have seen, much of this complexity is
associated with the elaboration of a separate and distinct head,
limbs and sense organs through the activities of neural crest
cells. It is thought that the origin of vertebrates from their
immediate invertebrate ancestors – nowadays represented by
lancelets and sea squirts – was accompanied by duplication of
large parts of the genome. This is borne out by recent research
on the sea squirt Ciona intestinalis, the closest invertebrate relative of vertebrates whose genome has been examined in any
detail. This creature has a genome of just under 16,000 genes,
comparable to that of flies or roundworms.[12]

The transition from invertebrate to vertebrate is, however,
much less marked than one might expect in terms of increase
in gene number. The human genome contains around 32,000
genes (although estimates before its sequencing were inclined

to put it higher, as much as 100,000). The genome of the mouse *Mus musculus* contains some 30,000 genes, 99 per cent of which have precise, one-to-one correspondents in the human genome – which is only to be expected, given that mice and men are both mammals, and have biochemistries and physiologies so similar that drugs intended for human use are routinely tested on rodents. The question here is not about the relationship between genomes and complexity, but about how it is that two animals with such similar genomes could come to look so different. The answer lies mainly in differences in gene regulation – in other words, in the function of the genome, rather than its structure. But some of the difference stems from the very different evolutionary histories of mice and men: products of the traditions, stretching back millions of years, in which primates and rodents have lived their lives – which have, in turn, shaped their respective genomic landscapes to produce the creatures we see today.

The most recent common ancestor of mice and men lived around 100 million years ago, a small, probably nocturnal animal which spent its time keeping out of the way of dinosaurs. The primate and rodent stocks were each well established before the dinosaurs became extinct, although the first primates and rodents would not have looked too different from each other. A hundred million years later, the external differences between mice and men are obvious to a proverbial degree. The genome of the modern house mouse, *Mus musculus*, is some 14 per cent smaller than that of humans, and the amount of junk DNA is also rather less. This does not mean that the mouse genome is any cleaner – only that it has evolved far more quickly than the human genome, giving the opportunity for mutations to destroy genes. LINEs, SINEs and other transposable elements remain active in the mouse genome in a way that has not been the case in the human genome for at least 40 million years.

Why this difference between mice and men? The best-laid plans of mice and men gang aft agley, it is said, but each in its own particular idiom. The answer could lie in the particular histories of rodents and primates, as well as in their present habits. The rodents are perhaps the most successful of all mammals, in terms of numbers of species and individuals. There are more species of rodent than of all other mammals put together, and rodents remain one of the few groups of mammals in which new species are still evolving rapidly. Rodents do not live for very long and reproduce profligately, and there is no reason to think that they have ever followed any other lifestyle. This means that at any given time, many millions of rodent genomes exist, and the rapidity of rodent reproduction means that these genomes are constantly being shuffled around. Large populations of active genomes are fertile ground for mutation and for the spread of transposable elements of all kinds.

Primates, the group of mammals to which we belong, are rather different. Unlike the abundant rodents, primates have always been rather rare creatures, specialized for life in forests. They tend to live in small populations, have relatively small numbers of offspring and live for a long time. This tendency has been sharpened with the evolution of large primates such as apes, and the ancestors of humans, in the past 20 million years or so. It is a fact of life that larger mammals live longer than smaller ones, breed less often and have fewer offspring. Genomes in historically small populations of rare, long-lived creatures offer few prospects for the spread of genomic parasites such as LINEs and SINEs, which explains why the junk in human DNA is static compared with that of mice. *Sahelanthropus tchadensis*, the earliest known member of the human lineage,[13] lived between 6 and 7 million years ago – coincident with the last known activity of transposable elements in the genome.

More interesting than the slow transformation of the human

genome into its own mausoleum is how it has come to be that such a vitiated product of evolution became able, within the past few tens of thousands of years, to make a transition that is, as far as we know, unprecedented in the entire history of life – the transition into self-awareness. Human beings can now study themselves and their origin, whereas there is no sign of any such ability in any other organism. If the human genome really holds the key to what it means to be human, signs of this transition should be abundant in the human genome, and exclusive to it. But there are no such signs. Indeed, the human genome is qualitatively no different from that of any other organism. Yes, it produces human beings with great accuracy, and does not produce mice, ostriches or even tuberculosis bacteria, but we can see in the human genome no obvious clues as to why this might be so.

Various attempts have been made to find direct, one-to-one correlations between certain genes and identifiable features particularly associated with our humanity, such as language or the acquisition of large brains. One of the more engaging examples concerns a human regulatory gene designated *FOXP2*.[14] This gene is found widely in mammals, and varies little from one species to another – with the exception of humans, in which a small base change not found in other mammals may have resulted in an important change in the properties of the protein encoded by *FOXP2*, potentially creating a site susceptible to a biochemical change called phosphorylation. This is one of the most important biochemical reactions in the body, widely used as a way to activate proteins. In short, phosphorylation is a kind of on/off switch. It could be that a small change in the human *FOXP2* gene supercharged what had hitherto been a rather quiescent protein, allowing it to do new things, perhaps create new evolutionary possibilities.

All normal humans studied so far – and no animals, not

even chimpanzees, our closest relative — have this distinctive, potentially hot-rodded form of *FOXP2*. Rare humans with mutations in the gene may have a syndrome in which they have difficulty coordinating facial movements with the finesse required to articulate spoken language. Taken together, the work on *FOXP2* suggests that at some time in the past few million years a mutation in *FOXP2* arose in the human lineage which might have allowed for refined coordination of facial musculature. The application of this mutation to the evolution of language offered such an enormous selective advantage to those humans who bore it that it quickly spread throughout the human gene pool. That the mutation is advantageous can be judged by the fact that animals without the mutation are mute — and that people with other mutations in *FOXP2* are not as articulate as their fellows.

That makes for a great story, but it shares the same flaw that bedevils all studies that focus on single genes: there is no simple connection between the proximate, biochemical function of the gene and the trait with which it might be associated. Nobody knows precisely what *FOXP2* does, and even if they did they would still be left wondering why this gene happens to be associated with the coordination of facial musculature, rather than some other trait. So, in the absence of a clear path between gene and trait, we are forced to construct a kind of fable about the evolution of language in our ancestors. Bateson might have reacted to this tale with as much disgust as to the 'stories' about vertebrate evolution that spurred him into writing *Materials*.

The fact that the human *FOXP2* protein might be phosphorylated in a unique way suggests something of potential significance: the creation of a new link in the network of genes whose activity makes us human, as distinct from mice or chimpanzees. On the other hand, people with mutations in the

FOXP2 gene are not suddenly deprived of their humanity. After all, parrots and voicemails can talk, but that doesn't make them human. What is important, therefore, is not the gene but the network, in which *FOXP2* plays its part alongside every other gene in the genome. The prize of identifying the connection between genome and humanity would be huge, provided we can get away from the headline-grabbing approach in which single genes are confused with single traits, and instead study whole suites of features which together define our humanity.

Perhaps the single most important such suite of features that determines our humanity is our long childhood – a trait so defining of humanity, and yet so complex that its evolution could conceivably have involved or affected every gene in the genome. The lives of most animals are easily told – they are born, they grow, they have sex and they die. Humans do things rather differently. Unlike chimpanzees, human babies are born in a relatively unformed state, and this is especially true of the brain. As a result, humans spend a disproportionate amount of their lives learning and growing. As if to match this, we tend to live for many years after we have ceased reproducing, leading to a cadre of elders, especially grandmothers. It is now well known that parents have a higher chance of seeing their children thrive if their own mothers are there to help.

This is most clearly seen in traditional societies in which day-to-day survival is more of an issue than in the developed world. But even in the affluent West, it is suspected that women who delay starting a family, and who have fewer children, tend to be those who live the longest and are therefore best placed to give their own children a good start in life – especially if they have a mother, or elderly relative, to baby-sit during board meetings. The very existence of grandmothering places a selective value on long life, and will in turn promote longer childhoods and adolescence. This, in turn, promotes late reproduction, long post-reproductive life, longer lifespan, and so on, in a perpetual loop of mutual reinforcement which may also involve the further growth of the brain, changes to the anatomy of the skull, face, hands and other parts of the body, the evolution and timing of puberty and the menopause, the growth of human family structures, societies, the evolution of art and culture, and perhaps many other aspects of our lives.[15] The evolution of language, and changes in genes such as *FOXP2*, must be considered as part of what one might call the Humanity Syndrome, a greater syndrome of humanity connected with the evolution of childhood. Mice, in contrast, are born in a relatively advanced state, reach sexual maturity very quickly, breed vigorously, and have short lives during which they exchange no spoken words at all. But mice have neither teenagers nor grandmothers: nor have they art, culture, spoken or written language, warfare, heaven or hell.

To discover when, in evolution, the Humanity Syndrome first emerged, you need look no further than your teeth. Thanks to the assiduous monitoring of children's teeth by thousands of dentists over decades, the development of any one individual can be judged very accurately by the state of their teeth. Given that most of the fossil record of the human lineage consists of jaws and teeth, scientists have been able to project dentistry back into the past, using fossil teeth to judge the development of our extinct relatives – and learn whether or not they had the long childhoods characteristic of humans.[16] As far as we can tell, the earliest members of the human family capable of making tools lived in Africa about 2.5 million years ago. Despite their technological skills, their life stories – to judge from their teeth – would have been very little different from those of chimpanzees living today. *Homo erectus*, which emerged in Africa some 2 million years ago, was the first member of the human family to have the tall, upright carriage we associate with humans rather than apes. *H. erectus* learned how to tame fire, made tools of breathtakingly fine craftsmanship, and may even have voyaged across the open sea. But even this creature lived out the life of an ape — to look into the eyes of *H. erectus* would have been to see a consciousness no more developed than that of any savannah predator, such as a lion or hyena.[17] The childhood of *H. erectus* was inhumanly brief.

What we now understand as Humanity Syndrome evolved in the common ancestor of modern humans and our closest extinct relative, *H. neanderthalensis.* We know that Neanderthals were both highly intelligent and technologically capable. The evidence of deliberate burial, as well as the remains of old and crippled individuals, is circumstantial evidence for the importance of post-reproductive elders — grandmothers — in Neanderthal society. But we do not know if they could talk, or whether they had those characteristically human mutations in *FOXP2*. But even if we had whole Neanderthal genomes available for study, it would still not be possible to point to any sequence of DNA and say, yes, this is the spark of humanity, the seat of the soul. We have a good idea of when the Humanity Syndrome arose, but we have not the remotest clue about how it relates to the raw DNA sequence of the genome.

Studying single genes is no help. What we need is to be able to create a model of the interactions of all the genes in the human genome — something akin to the model of the segmentation module created and studied by Garrett Odell's group at Seattle. Only then will we have the script to the human drama in our hands, such that we will be able to tell of antres vast and deserts idle, of the cannibals that each other eat, of the anthropophagi, and of men whose heads do grow beneath their shoulders.

12

My travels' history

We have come a long way since Aristotle first wondered why it was that pregnant women did not menstruate, and concluded that babies were made from unshed blood. Aristotle is rightly regarded as the father of biological science, and the history of biology can be seen as a great quest to elaborate on Aristotle's hypothesis, to understand how babies are made and how it is that all the diverse creatures of the natural world can emerge from tiny, formless eggs – *Ex Ovo, Omnia*. Cast as such, the story of biology can be told in an unbroken skein, through Harvey and the preformationists who, finding Harvey's ideas incomplete, created a robust theoretical edifice which stood for longer than the entire history of genetics has so far endured; to the nature-philosophers, who turned the tables yet again, and without whose idealism Darwin's ideas could not have come into being in the way they did; to Bateson and Morgan, reacting against what they saw as Darwinism's deficiencies, and setting in train the amazing discoveries of experimental genetics that have marked the past century; right to the present generation of computational biologists whose concepts of the genome as a network seem to resolve the questions that Aristotle asked.

The network concept does seem to embrace all the qualities of the genome I have discussed throughout this book. As a human being – and particularly as a parent – I am as amazed as Aristotle must have been that creatures as intricate and delicate as babies are produced so routinely and with such fidelity. This reliability is a consequence of a genome constructed as a network, which can produce a human being in the most varied and trying of circumstances, at the cost of a small amount of variability in the finished result that we accept as the normal range of variation. As such, the reliability of networks provides, almost as a by-product, an explanation for the existence of variation, a puzzle which perplexed Darwin and Bateson. However, the existence of this variation is something for which we should be grateful, for were the genome to conduct itself with the precision of an engineer, and not tolerating a certain amount of woolliness in its operation, every single variation might be monstrous and no human beings would ever be born.

Such woolliness explains why it is that in its embryonic development every new baby passes through stages reminiscent of the evolutionary history of the species. Babies are not designed from scratch after each conception, but are built according to networks which already exist: the networks that built their parents, their grandparents and all their other ancestors right back to the beginning. These networks were repeatedly damaged and patched up, and accumulated small changes, some of which not just created new individuals but provided the germ of variation that created new species. Change was built into networks which, thanks to their forgiving nature, functioned none the less – building up traces of their histories as they went along. The scars of evolution that mark the genome are matched by those that started with the division of one cell into two identical daughters, and so, after billions of years, created the exquisite marvel that is embryonic development, in which, for example, a human baby goes through a disc-like stage, just like a modern egg-laying reptile, even though the last member of the human lineage that laid eggs lived more than 100 million years ago. But these, and other reminiscences – of lampreys, and sea squirts, and other wonders – are just that: reminiscences, not an orderly parade that might be used to plot evolutionary history with any great accuracy. They are impressions, cartoons, doodles on the chalkboard of life which have not been fully erased.

If networks allow for such laxity, why are there so many distinct species in the world, and how do we know that, say, a red-cheeked bulbul is an entirely different thing from a Western tanager – or, for that matter, that an individual fly of the species *Drosophila melanogaster* is quite different from a fly belonging to the species *D. simulans?* The question of why species are, in general, so distinct from one another has long been perplexing. Networks are so forgiving, yet the world's creatures are not all members of one huge, heterogeneous, continuously variable continuum. To put it in more concrete terms, human beings come in a wide variety of shapes and sizes, but they are all equally human, and the world is not full of quasi-human creatures that are more or less like chimpanzees or other creatures. This, in essence, is the demand that Bateson made of the biometricians: if variation within a species was limited to extremely small gradations of traits, as the biometricians insisted, why were differences between species so clear to see?

I suggest that the network concept of the genome might provide the answer that Bateson sought. Although the output of a network can vary, the pattern of connections that comprises the network does not actually change within a given species. All humans, for example, are products of the same network, whose overall output depends on how the various genes and modules are wired together. As Garrett Odell and his team at

Seattle showed with the segment-polarity module, segmentation can be produced in an enormous range of situations, provided the network itself is wired up correctly. Even if genes in the network were damaged beyond repair, as in the fruit-fly monsters created by Nüsslein-Volhard and Wieschaus, segments might still be produced.

But what if a mutation resulted in a change in the pattern of connection itself? The change need not be very great: for example, a mutation that created an operator sequence where none existed before might allow repressors or other factors to bind there. The result would be the creation of a network qualitatively different from the one that had existed. As the Seattle group showed, networks are very sensitive to such changes in connectedness. If they are not connected correctly, they are very likely not to work at all. Any such change would therefore have to be small, and perhaps not initially of much significance. After all, the creation of a new operator sequence would be of little consequence if no repressor yet existed to bind to it, or if such substances that might do so were better at binding elsewhere. However, the result of such a change would be to introduce an element of instability into the network. It is a property of networks, and distributed systems, that they are inherently good at working round such problems. Mutations that introduce changes in the connection pattern in a network might soon be followed by a host of other small, compensatory changes in quick succession to give the new network some kind of equilibrium. Given the right circumstances, such changes might form the basis of a new species.

This idea, that networks might switch rapidly from one so-called 'metastable' state to another, is nothing new in science, especially in physics and mathematics. To apply it to the genome might seem speculative in the extreme. Nevertheless, the idea does seem to fit with what we currently know about speciation –

that is, the genetics of the origin of species. Speciation is all about sex and reproduction. It can happen only if the creatures in one population stop breeding with those in another, for if they continued to do so their genes would stay forever mixed, divergence could not happen, and we would indeed all be part of a single, heterogeneous continuum with people at one end, flies at the other, and the ghost of Franz Kafka somewhere in the middle. In the real world, however, creatures go to enormous lengths to avoid misalliance, especially if a pair of different creatures look like each other. In groups of animals in which there are many distinct but very similar species, males or females evolve very exaggerated and specific sexual features to ensure that members of one species do not mistakenly mate with members of another. Many species of spider, for example, are identical but for the elaborate genitalia of the male which, like a key, will only fit into the genitalia of the female of the same species. Female fruit flies will swoon only before males who perform a specific mating ritual with the requisite precision. If the process of speciation depends on genetic variation, the genes involved will be associated with traits connected with sexual anatomy and behaviour. Theories of speciation based on collections of genes – rather than their genomic totality – are good at explaining how divergence proceeds once it has begun, but are less successful at describing the very first changes that lead to divergence. And the still highly controversial question remains – whether speciation consists of many small changes in a large suite of genes, or a smaller number of changes in a few, key genes.

I think this might be the wrong question, because it concentrates too closely on the behaviour of genes in isolation instead of looking at the behaviour of the whole genome. I suspect that speciation might involve changes in the interaction patterns of genes or modules associated with sexual behaviour. Because all modules are connected to some degree, these genes or modules would have close links with other genes or modules specifying other body parts or functions. Because sexual isolation is crucial to speciation, it is likely that such links are close: changes in, say, networks specifying sexual anatomy would have rapid knock-on effects on other parts of the body. Changes involving sex genes would have a domino effect, opening the way for natural selection to prompt compensatory changes in other, more remote parts of the network. Such changes of wiring might have the effect of amplifying any small, external change in behaviour, as well as any accompanying changes in form or function. Again through natural selection, these changes would feed back on the genome, providing the opportunity for many other changes to occur. Eventually the network would achieve a new stability, a new equilibrium – and a new species.

If this seems speculative, it is – and deliberately so. But in fact, a tentative model of such connections already exists and has already been mentioned – the case of the fruit-fly gene *bric-a-brac*. This gene is responsible for the gender-specific pigmentation of fruit-fly abdomens, which has a role in the flies' sexual behaviour and choice of mate. At the same time the gene is part of a regulatory network – involving the *Hox* genes – that shapes the abdomen itself, which suggests a connection between the courtship behaviour of flies and the shape of their bodies.[1] Once a new species is formed, perhaps in the wake of one small change, other compensatory changes might soon follow so that it becomes progressively more difficult for a member of a new species to mate with a member of the ancestral species and produce viable offspring. As species accumulate changes immediately after the split, hybrids first become weak, then sterile mules, then inviable monsters that die in the womb, then cease to exist.

All of the above is advanced in the spirit of tentative speculation, but it does seem that the network view of the genome has the potential to explain an enormous amount about its past history and current behaviour. It is certainly a better model of reality than the clichéd description of DNA as a 'blueprint'. However, we must guard against proclaiming that the network view is the final and definitive answer to how form is created from the formless. If the history of scientific enquiry teaches us one thing, it is that we should never assume that progress is either uniform or proceeds in one direction, such that our predecessors were always wrong, and the further we delve into the past the more wrong we find they were. The history of biology, told here as the search for the agency that created form from the formless, shows that such a patronizing view of the past is in fact worse than wrong, for it blinds us to the work of great minds, all of whom made important contributions to the way we think now. Like the genome and the process of embryonic development it engenders, the development of our own thought bears the scars of its own evolution without which it might be very different – perhaps incomprehensibly so. Either that, or it might not exist at all. We owe a debt of respect to our predecessors, even though their thoughts might seem to us misguided or even bizarre.

The story of how our genome came to be understood can be seen as a simple quest. Alternatively, it may be seen as having been prey to successive waves of intellectual revolution. As Harvey's experiments overturned centuries of orthodoxy based on Aristotle, the work of the preformationists overthrew Harvey's epigenesis, and preformation succumbed in its turn to a resurgent epigenesis that became nature-philosophy, which produced what would eventually become modern embryology. In the meantime, Darwin's work was a reaction to the idealism of nature-philosophy in which variation was seen as a hindrance to understanding, rather than a phenomenon demanding explanation in its own right. Unable to achieve a complete understanding of the sources of variation, however, Darwinism was eclipsed, if not overturned, by the new genetics of Bateson and Morgan, and was not rehabilitated until these early geneticists had all but left the scene.

That rehabilitation, by Morgan's student Dobzhansky and others, created population genetics – an understanding of evolution based on collections of genes pictured as individual agents, often in competition with one another. This world-view is perhaps most eloquently expressed by Richard Dawkins in *The Selfish Gene*, and has successfully explained much about the way nature works. But a new view is now emerging which claims much more. The network view offers fresh approaches to enduring problems such as speciation, which population genetics has found very hard to explain fully. In addition, the new view suggests opportunities for discussing the grand sweep of evolution and the minutiae of individual development within the same theoretical framework – something that population genetics has very largely failed to achieve.

One of the great strengths of population genetics is its universality. Evolution is discussed in terms of the ebb and flow of different genetic variants, or alleles, among populations of individuals. When put so simply, the findings of population genetics can easily be expressed in mathematical terms, allowing scientists to devise predictive theories of evolution of disarming elegance and enormous power. The equations of population genetics are not tied to the life or history of any particular organism, and are therefore universal. Among many other things, they can explain why bees are so faithful to their queen; why female birds in apparently monogamous relationships cuckold their mates with such abandon; how mice discriminate between their own kin and a potential mating partner; and how cannibalistic salamanders select their next victim. I fully expect that when we encounter life forms on other planets, the social behaviour of aliens a dozen light years away should be expressible by the same formulae that can be used to summarize the courtship of two flies on a cow pat in the field next door.

Curiously, it is in this very universality that the weakness of population genetics lies. Because it can apply to any population of organisms, anywhere and at any time, it is not very good at expressing the particular circumstances of the evolutionary history of which we are a part. In particular, it is impossible to extrapolate from the behaviour of genes in populations to the evolution of species over geological time, because one can never assume that the conditions in which organisms are evolving in one set of circumstances will necessarily apply in a different situation millions of years later. As soon as this assumption is made, however, one strays into the realm of fable – meeting stories in which, for example, an overwhelming and unswerving selective need, maintained over millions of years in all circumstances, drove a population of reptiles to evolve feathers and become birds, or coerced the ancestors of humans to adopt an erect posture, freeing their hands for carrying food and nursing babies. Such tales are neither better nor worse than Lamarckian *besoin* or nature-philosophical urges towards cosmic perfection, and Bateson – rightly – found them deplorable. Modern Creationists have been very quick to exploit this tendency for population genetics to be extrapolated into realms that it was not designed to address. Creationists will accept the 'microevolution' of adaptation within populations, while denying that the natural world above this relatively trivial level need be explained by a kind of 'macroevolution', especially as population geneticists have failed to describe macroevolution convincingly.

None of this is meant to imply that the population genetics that explains so much about the behaviour and constitutions of animals and plants, and has served us so well for nearly a century, is either misleading or wrong. However, it may very well be that some aspects of evolution, such as speciation and macroevolution, might be answered more successfully by taking a more rounded approach – that is, with reference to the behaviour of the genome as an integrated entity, rather than as a collection of genes: as a genome whose complexion is sensitive to its own history. It is possible that the network view as I have sketched it here has the potential to become a more complete, inclusive description of reality in which population genetics describes microevolution – that is, relatively small genetic changes and events within species – better than the more dramatic descriptions based on macroevolution, the appreciable changes of form over measurable intervals of geological time. By way of analogy, everyone recognizes that Einstein's relativity offers a better account of gravity than Newton's laws, because it encompasses, within the same theoretical framework, both everyday events and the more extreme phenomena for which Newtonian descriptions are inadequate, such as the motion of massive objects travelling close to the speed of light. In most circumstances, however, Newtonian rules will apply: despite relativity, apples still fall from trees the way they always did.

More fundamentally, however, the network view could represent a reaction to the reductionist tendency of much in modern science that seeks to apprehend structures in terms of their component parts, rather than as complete entities. In contrast, a view that connects the individual development of a human baby with the evolution of all life over billions of years, linking the microcosm with the macrocosm in true nature-philosophic style, is something Goethe might have appreciated. Only time will tell whether the network view will be seen as something set in opposition to more reductionist styles of genetics, in the way that Bateson's genetics was a reaction to Darwinian evolution before their eventual reconciliation; or that preformation was couched in opposition to epigenesis, before it was discovered – in the relationship between the genome and embryonic development – that there is merit in both views.

I rather hope that, at least to begin with, the network view of the genome will come to be seen as something opposed to the reductionist programme, because it is otherwise impossible to grasp the scientific implications of the emergence of humanity if the human genome is considered simply as a collection of genes, or a list of bases, rather than an entity with a unique history and unitary properties all its own, properties that cannot be predicted from single genes or sequences of DNA. After all, we cannot point to any one sequence in the human genome and find a homunculus, nor will we ever be able to spot, in a forest of sequence differences, those that led to the spark of sentience. The differences are everywhere, and yet they are nowhere consequences of the function of the network as a whole, not of any one of its parts that we can identify.

In which case, to state that the draft sequence of DNA will tell us what it means to be human is to overstate the case by a large margin. We have come a long way since Aristotle, but we have a great deal left to do, and to learn. The great adventure has only just begun.

13

Jacob's ladder

Life on Earth has arrived at a threshold. After more than 3 billion years of evolution, the genome has, in the past few thousand years, wired itself in just the right, reflexive way to create creatures capable of wondering about the emergence of form from the formless. In just the past few decades, these creatures have started to acquire detailed knowledge about how this process actually works – and how it might be changed. The chances that this point should have been reached right *now*, while you are reading this book, seem infinitesimal, given the scale of the Universe in time and space. However, it is legitimate to ask whether the incredible length of these odds might not be more imaginary than real: it could be that 3 billion years of evolution really are required for self-awareness to emerge on any planet, given that such a property might demand genomes of more than a certain amount of complexity.

On the other hand, the increase in complexity is neither linear nor steadily upward. In terms of gene numbers, at least, genomes have increased dramatically in size only three times in the history of life, discounting events surrounding the origin of life itself (of which we know next to nothing). These increases are connected with obvious transitions in the external state and habits of organisms. The first increase happened more than a billion years ago, with the first appearance of integrated, eukaryote cells (ones with nuclei). This was by far the largest single increase in gene number, from around 1,000 to 4,000. The second event – a further increase to around 10,000–15,000 genes – was connected with the invention of organized, multicellular creatures, particularly animals, around 600 million years ago. The third event saw one or more stepwise increases in gene number, to around 30,000, and was associated with the evolution of vertebrates some time before 550 million years ago.

Within these categories, however, gene number says little about complexity, still less about the evolution of human sentience: human beings do not have significantly more genes than do other vertebrates, such as mice or puffer fish. Conversely, were complexity simply a matter of numbers, self-aware organisms could have evolved at any time since vertebrates first evolved. And we could go further: our ignorance of the relationship between internal genomic complexity and external traits prevents us from claiming, beyond doubt, that there is something special about the size of vertebrate genomes (rather than, say, bacterial genomes, or invertebrate genomes) that predisposes them to evolve self-aware creatures, such that smaller genomes would invariably result in, for want of a better term, dumb animals.

So, given the present state of knowledge, we cannot know for sure that sentience cannot arise in creatures with far fewer genes than are routinely found in the genomes of vertebrates. Self-awareness could therefore have arisen at any moment in the past 3 billion years – which makes it all the more special that it has happened to emerge, on the now.

It should now be apparent that there is no direct relationship between gene number and complexity. Sentience lies not in an increase in the number of genes, but in a qualitative change in how genes are organized into regulatory networks. And because each network has a history, and is not created anew in each generation, it cannot be said that the transition to sentience was equally probable or possible in any one of the tens of millions of different networks that existed at any given moment in the history of life, irrespective of the number of genes they contained. The change happened just once, by virtue of a small change in the connectedness of the genomic network peculiar to the immediate ancestor of modern humans. That network, because of its particular history and connection pattern, happened to be sensitive to whatever small change it was that occurred, such that larger changes immediately followed – including the transition to self-awareness.[1]

The change need not have been very great: a mutation that created an operator sequence where none before existed, or perhaps – as in *FOXP2* – a mutation that created a new region in the protein with the potential for chemical activation by an enzyme. Once made, the change would have been followed by a flurry of other changes, so that several million years after the event it might be very hard to spot precisely what the original change was, even were we able to compare the complete genomic networks that specify humans with those of closely related, non-sentient species.

Our closest extant evolutionary relative is the chimpanzee (*Pan troglodytes*), whose genome sequence is being prepared as I write (late 2003). Comparison of the human and chimpanzee sequences will yield much of great interest, but the secrets of humanity are not likely to be among the immediate benefits. In fact, the chimpanzee and human genomes will look so much alike in detail that the degree of physical difference between the two species will be hard to credit.

The answer will surely lie in a concerted approach – working out those genes, and those modules, that are related to individual development. Chimp adults look very different from human adults, but we look much more similar as babies, foetuses and embryos. Small changes during early development may translate into larger ones later on. However, any changes that we find need not mark the *causes* of differences between the two species – the 'smoking gun' of humanity – but could instead represent the *consequences* of a whole host of subtle changes in development, sexual behaviour and life-history evolution of the kind I discussed in Chapter 12, all precipitated by an unknown number of genetic changes that will be extremely hard to spot. After all, one small change in a DNA sequence looks very much like another, and it is estimated that *18 million* base-pair differences separate the human genome from the last common ancestor of chimps and humans, quite apart from any differences that might be peculiar to the ancestry of chimps.[2] Not one of these changes will be specially marked out for our benefit in flashing neon lights as the one, crucial change from which all humanity must flow.

In any case, geneticists will be less interested in extracting meaning from small-scale base-sequence comparisons than in looking for those changes that could have implications for how genes regulate one another. Large-scale alterations in gene order or chromosome structure are known to affect regulation, because they split up operon–like clusters of genes, forcing once widely separated genes into closer proximity. Even though the chimp sequence is not complete, we already know that long sections of at least nine chimp chromosomes are inverted (that is, turned back to front) relative to the same sections in the equivalent chromosomes of humans. Such inversions might be related to significant regulatory changes. In addition, one of the human chromosomes is known to have been created from the fusion of two shorter ones that are found as separate chromosomes in chimpanzees and other apes. This last change is likely to have occurred in the ancestry of humans, after our lineage split from that of the chimpanzees at least 7 million years ago. Many of the other changes, however, could be peculiar to chimps and have nothing to do with the human condition. It is worth noting that chimps have been evolving away from humans for precisely the same interval that we have been evolving away from chimps, and their genomes will have had the same chance as ours to accumulate change.

In the final analysis, it could be that any further questioning along these lines is fruitless, in that no individual detail of the chimp–human genome comparison is ever likely to reveal the genomic basis for that elusive quality we know as human. On the other hand, a complete description of *any* genomic network, particularly for a multicellular animal, would allow unprecedented insight into how form is created from the formless. Efforts in this direction are already under way. Once we have learned to describe one species in terms of its characteristic genomic network, others will soon follow. The comparison of the properties and connectedness of the networks of different species will enable us to explore the relationship between the general properties of networks and the forms of animals and plants. This will then give us the language we need to articulate fully the question of why there are so many different species of animals on the planet, rather than fewer or still more species; and why the species we see have adopted the forms they have.

A complete description of the genomic network for humans might not expose the seat of the soul, but it might open the way to a new kind of exploration whose prizes could be of incalculable worth – and which could also create immense potential dangers. For it is our lot as humans to be curious, and to be the first creatures to have evolved, as far as we know, the ability to make a conscious decision to initiate radical changes to our own genetic constitutions. We will soon be able to change the human genomic network.

Human-induced changes to the human genome are nothing new. The invention of agriculture and a sedentary lifestyle around 10,000 years ago led to stresses on the human frame that have measurably altered the human genome. The first farmers were smaller, less healthy and more prone to disease than their immediate forebears, who were hunter-gatherers. Various parts of the human genome show signs of natural selection in response to diseases, some of which did not exist before humans domesticated animals and started to live in close proximity to them. These diseases include plague (carried by the bacterium *Yersinia pestis*) and tuberculosis, both of which originated as diseases of animals and which have subsequently had a profound influence on human history. Empires have fallen and history altered thanks to smallpox, syphilis, measles and typhus. More recent instances of this kind include the influenza that came from poultry and killed more people in 1918–19 than in all the previous four years of war combined. The recent epidemic of the flu-like disease Severe Acute Respiratory Syndrome (SARS) is also believed to have come from poultry, and the ongoing epidemic of HIV-1 may have been transferred to humans through the consumption of primates in West Africa. An epidemic of Type II (non-insulin-dependent) diabetes is currently sweeping populous nations in Asia, with the potential to cause radical genetic change in the next century or so.[3] All these changes will have caused shifts in the frequency of different varieties of genes (alleles) in the human population – but none, as far as we know, has altered the fundamental genomic network that is common to all humanity. Furthermore, none of these changes, large and occasionally catastrophic as they have been, were achieved through conscious human action.

The first deliberate attempts to change the human genome

have been made in the past few years, in the form of gene therapy. This is an experimental medical intervention in which people with syndromes resulting from known defects in particular genes are treated with synthetic versions of functional genes – administered by infecting the patient with genetically modified viruses – in an effort to alleviate their particular condition. Gene therapy has been of limited success so far, and it is possible that it will never be applicable to more than a few syndromes, and even then only in specific circumstances in which no other treatment is available. Neither does gene therapy constitute an attempt to alter the human genomic network in any fundamental way. Indeed, nothing could be further from the minds of clinicians, who are seeking only to repair defects in the existing network, not to create new ones.

The urge to create, rather than simply repair, will come with developments in computational biology, perhaps along the lines pursued by the Seattle group, only very much larger and more complex. Given that it is already possible to simulate the activities of limited networks of genes in a computer, it is therefore possible in principle to imagine a digital description of any such module, or even the complete network that specifies the development and maintenance of an organism such as the fruit fly, or a mouse, or even a human being. That is, it should be possible to write a computer program that recreates human development entirely within a computer memory.

This approach might have great benefits for medicine. It might allow scientists and clinicians to model the formation of various organs entirely by computer, realizing them in laboratory conditions, and using the results to heal people born with various defects or who have suffered amputation or organ failure later in life. The network approach could grow new limbs for the limbless, create new skin for burns victims and give eyesight to the blind. It is also possible that the network approach can be used to alleviate the symptoms of genetic diseases, whether caused by single-gene defects (cystic fibrosis, phenylketonuria); more subtle, possibly regulatory interactions (insulin-dependent diabetes, heart disease, Alzheimer's disease and other forms of neurodegenerative disorder); or even wholesale gain or loss of entire chromosomal regions (Down's syndrome, in which patients have an extra copy of Chromosome 21). In a sense, though, this therapy is no different from present-day single-gene therapy in its motivation: to heal a defective network, but not to change it.

I predict that the first efforts to change the network will be driven by less high-minded ideals than the relief of suffering. Cosmetic surgery provides an instructive model for what might happen. Originally developed to alleviate the pain and dis-figurement suffered by burns victims, it was then applied to less life-threatening problems such as the removal of birthmarks. Although cosmetic surgery is still used to remove suffering, it is perhaps best known as an instrument for the elective fulfil-ment of desire – to have a more attractive body than the one offered by nature, or even a body of a different gender. It could be that the modification of genomic networks might also become an instrument of vanity, but one that offers far more potential for change than cosmetic surgery – which does not, after all, alter genes or networks.

At first, people might use network modification for rather predictable ends, for example to alter their metabolism to allow them to eat more without gaining weight, or to drink alcohol or consume drugs without ill effect (or detection by the authori-ties); to improve muscle tone, correct wayward eyesight, improve resistance to common infections (especially sexually transmitted ones), or change the size of breasts or genitalia. Criminals will enjoy being able to alter their faces, although it might be very much harder to morph the face of one person

245

into the likeness of another, or to alter fingerprints. DNA fingerprints – in contrast to the prints actually on your fingertips – are unique personal markers regularly used today in forensics. They are features of highly repetitive, junk DNA, and might not be affected by any kind of network modification. Given that network analysis will not reveal the secret of humanity, it might be particularly hard to effect changes in behaviour which might, for example, improve intelligence, remove criminal tendencies, or alter personality in ways that cannot already be achieved by using drugs such as antidepressants.

Once network modification has been widely adopted for therapeutic and recreational use, it might be used in more radical ways, reflecting tendencies in fashion or even politics. Solidarity with an oppressed minority – or allegiance to a favourite sports team – might be expressed by wearing skin of a different colour, for a day or a season. Military researchers and technologically minded terrorists might find a way for people to endure harsh conditions for long periods, assume unusual athletic abilities, see in the dark or even spit poison. Changing sex, adding new limbs or even growing feathers might reflect sincere personal choices, with religious or spiritual significance.

None of these changes need be permanent, so that once they have been made, the person need not live with them throughout life, nor pass them on to the next generation. Gene fixes need be no more permanent than, say, a tattoo is today. The subject of germline modification, on the other hand, in which parents effectively choose the attributes of their children, is a controversial one – as scientist and futurologist Gregory Stock found when he organized a conference on the subject.[4] *In vitro* fertilization (IVF) technology is gaining in popularity, especially among people who choose to wait until their thirties and forties before starting a family, when the risks posed by infertility and genetic disorder are higher than for couples who start younger.

People who delay reproduction are also likely to be educated, wealthy and informed about the choices on offer, which now include the gender of their offspring, and screening embryos to ensure that a child is not born who suffers from any of a range of genetic disorders.

IVF and embryo screening – and even cloning, reports of which are still unconfirmed in humans – are examples of conscious intervention and choice in reproduction, but none of them entail or require any kind of genetic modification. It may become possible, however, to impose genetic change on unborn children. This might initially involve *in vitro* gene therapy to alleviate single-gene defects otherwise missed by screening, but genomic network modification might also become an issue. People might wish to use network modification to change the attributes of their children as a way of expressing their own desires or beliefs. Whatever the outcome, such interventions will raise formidable ethical questions, most notably whether parents have the right to impose fundamental and possibly indelible changes on the form and even the personalities of people who would be in no position to raise objections, nor make informed choices of their own.

Of course, there could be a loophole. Rather than imposing their designs on their direct, biological offspring, a parent might design a child entirely from scratch. If it is possible to create a computer model of the specific genome network of a human being, it might be possible to design humans with any desired trait, whose genome would not be constrained by parentage. The design might be synthesized as DNA, packaged into an egg and brought to term in an artificial womb. This implies the existence of technology far superior to anything we have at present, although the germs of it can be seen in technology used in IVF, cloning and the support of extremely premature infants.[5] The effects of such a strategy on human society can

247

hardly be imagined. Considered at the most superficial level, however, parents would not exercise a completely free choice, but would tend to select variants on a rather restricted range of designs on the basis of fashion. You could imagine a situation in which children in a particular year group would all tend to look like popular entertainers or sports stars of the day.

Somewhere along the line, these children, created entirely artificially, would acquire sentience by virtue of their construction according to the stock human genomic network. However, it is legitimate to wonder whether the computer representation of the human genomic network used to create these children might not itself acquire a semblance of sentience, as a result of the connectedness inherent in the program. Would a virtual human grow up in a computer – would it have a personality?

As we learn how to design, create and modify humans, we will do the same for many animals, plants and microorganisms, changing the world around us irrevocably, for good or ill. New lives, new organisms will be created to cater for our slightest whim, our every convenience. Solving the ethical questions posed by the potential to exercise this kind of power and control on the world around us will require a degree of detachment and maturity not evident in current debates about genetic modification or assisted reproduction. Genetic modification (GM) of crops is equivalent to gene therapy in humans, but the debate on the desirability of GM tends to entrench positions conditioned by, or in opposition to, business or political interests, rather than making progress through a detached consideration of the advantages or disadvantages of the technology to economies and markets. Reaction to any kind of assisted reproduction, conception or cloning reflects either parental selfishness or desperation, or the will of opponents to impose a narrow ethical view on others, whether or not this advice is welcome.

But GM and IVF are as nothing compared with the effects

of the genomic network modification that is to come, and the standard of ethical debate must rise to meet it. A novel problem raised specifically by genomic network modification – not evident in gene therapy, GM, IVF or cloning – is that we could be able to modify human networks in such a way that we might lose the indefinable quality of humanity that makes us special: that same edifice upon which our ethical, legal and moral codes all stand, and on which our lives and loves are based. To what extent will the products of modified networks be new species, inhuman, or even 'posthuman'?[6] What will their relationships with unmodified humans be like? Will they be our servants, our masters, both – or neither?

Many of these questions may seem overly speculative or fanciful; they are certainly far from our current needs. I would argue, however, that these technological changes could happen faster than we might think, and certainly much faster than some of us might like. There is an urgent need for debate on these issues, but not the kind of debate that acts only to amplify the religious, moral or political views of the present. Such views will be of no help in the sensible management of the novel challenges that the future will throw at us, and may actually obstruct clear thinking about what is best.

The greatest immediate challenge is posed by wilful public ignorance, cozened by woolly thinking, sentimentality and, above all, lazy journalism. It should now be clear that every method to assist reproduction or improve fertility that has so far been invented does nothing more than nurture processes that occur naturally. When a couple choose the sex of their baby, they are only selecting from a number of embryos that have already been created by natural processes – the same story of sperm-meets-egg that has gone on for billions of years. These embryos are genetically and genomically no different from any other human – yet journalists cannot seem to refrain from using headline language, such as 'designer babies', and referring luridly to *Frankenstein*. It occurs to me that if people really thought about the implications of designing a baby – from scratch, on a computer, gene by gene, connection by connection – their feelings of horror might be justified. If we are to secure the future for our children, we cannot afford the luxury of being mildly titillated at the expense of rational thought.

We must try to raise our game to imagine the unimaginable, with cool heads, and we must do it soon. We have to devise rational ways of thinking about situations that might arise in the future in which our descendants, accustomed to living in a world of widespread and accepted genetic change, might have to make moral decisions that we can scarcely even imagine. What will society be like in a century or so? Will we be able to grow extra limbs or even change sex as easily as changing clothes? Will 'designer babies' be a reality? What would be the effects of such changes – such implied power – on the things that really matter to us now: friendship, love, the whole business of two people setting up home under one roof?

Will our descendants, capable of altering everything we traditionally regard as definitive of humans, still think of themselves as human in any sense that we can now recognize? Will they become a higher order of being – or will they become like *Homo erectus*, a creature that walked like a man but thought like a wild beast? To have engineered a situation in which the brief spark of humanity currently resident on Earth winks out would be a tragedy as immense as the backdrop of time against which evolution is set.

In the Book of Genesis, God gave Jacob a vision of angels ascending to heaven and told him how his descendants would inherit the Earth. But does this licence extend to becoming angels ourselves? The transformation of human beings from apes into angels may sound like pure science fiction, something we needn't worry about any time soon. But the clouds are already gathering. Genetics was established as a discipline only a century ago, and now we have the draft sequence of the human genome. This sequence does not, in itself, tell us what it means to be human.

Now is the time to start learning what does.

Notes

CHAPTER 1

1. I owe much of the embryological detail described in this chapter to *Essentials of Human Embryology* by William J. Larsen (New York: Churchill Livingstone, 1998).
2. I discuss the evolution and development of sea squirts and lancelets, and their relationship to backboned animals, in my book *Before the Backbone* (London: Chapman & Hall 1996).
3. The earliest-known vertebrates are the fossils of fish-like creatures from China that lived some 530 million years ago (see the paper by De-Gan Shu, Hui-Lin Luo, S. Conway Morris, Xing-Liang Zhang, S.-X. Hu, Liang Chen, J. Han, Min Zhu and L.-Z. Chen, *Nature*, vol. 402, 1999, pp. 42–6). However, we have no way of knowing whether these particular fossils were our lineal ancestors.
4. Estimates of the numbers of species on Earth at present vary enormously, from around 1.4 million to 200 million.
5. For more about the lives of the remarkable Rothschilds, see *Dear Lord Rothschild: Birds, Butterflies, and History* by Miriam Rothschild (Glenside, Pennsylvania: Balaban, 1983) and *The English Rothschilds* by Richard Price Davis (Chapel Hill: University of North Carolina Press, 1983).

CHAPTER 2

1. The background to the early history of embryology as described in this and the next chapter, including preformationism and much else

of interest, can be found in *Early Theories of Sexual Generation* by F. J. Cole (Oxford: Clarendon Press, 1930), *A History of Embryology* by Joseph Needham (Cambridge: Cambridge University Press, 1934), *Investigations into Generation 1651–1828* by Elizabeth B. Gasking (London: Hutchinson, 1967) and a sparkling recent book, *The Ovary of Eve*, by the biologist, poet and novelist Clara Pinto-Correia (Chicago: University of Chicago Press, 1997).

2. Quoted in Needham, *op. cit.*, p. 48, and attributed to Charles Singer.
3. *Ibid.*, p. 49.
4. *Ibid.*, p. 50.
5. 'Cell theory' as we know it today was a product of the nineteenth century, but its first stirrings – in the idea that organisms were divisible into small units called 'cells' – lay the other side of the Restoration, with the publication in 1665 of *Micrographia* by the English micro-scopist Robert Hooke (1635–1703), who described the cells of cork wood.
6. For more on Harvey's insight see the essay 'Where do babies come from?' by R. V. Short, *Nature*, vol. 403, 2000, p. 705.
7. Gasking, *op. cit.*, p. 43.

CHAPTER 3

1. This is quoted on p. 2 of F. J. Cole's *Early Theories of Sexual Generation* (Oxford: Clarendon Press, 1930). Cole gives a date of 1678 for Huyg-ens's statement, but no attribution is given for Cole's translation, which he may have done himself.
2. The concept of 'fertilization', as distinct from fecundation, depends on the union of egg and sperm. Spallanzani did not believe that sperm had any role in generation, so the term 'fertilization', in the sense of generation, would have had no meaning for him.
3. Elizabeth B. Gasking, *Investigations into Generation 1651–1828* (London: Hutchinson, 1967), p. 132.

CHAPTER 4

1. Quoted by Elizabeth B. Gasking, directly from Wolff's *Theoria genera-tionis*, on p. 103 of *Investigations into Generation 1651–1828* (London: Hutchinson, 1967).
2. *Ibid.*
3. Robert J. Richards gives a good, brief account of the ideas of the

nature-philosophers in *The Meaning of Evolution* (Chicago: University of Chicago Press, 1992). The quote from Lorenz Oken is from this source (p. 39), translated from Oken's *Abriss des Systems der Biologie* (Göttingen: Vandenhoek und Ruprecht, 1805). Richards's account is given in the context of Darwin's early ideas of progressive evolution, before natural selection took shape.

4. The scientific works of Goethe are conveniently available in English, in *Goethe: The Collected Works, Volume 12 – Scientific Studies*, edited and translated by Douglas Miller (Princeton: Princeton University Press, 1988).

5. See an article by Günter Theissen and Heinz Saedler, *Nature*, vol. 409, 2001, pp. 469–71.

6. This idea is an important part of the thinking of the twentieth-century philosopher Rudolf Steiner (1861–1925), whose 'anthroposophy' draws heavily on the works of Goethe and on nature-philosophy in general. The preoccupation of anthroposophists with holism, the interrelationships between body and spirit, and general ecological concerns, can all find their roots in Goethe and in nature-philosophy more generally. Anthroposophists also promote some of Goethe's more outmoded concepts such as his theory of colour, concentrating on its aesthetic rather than its scientific value.

7. This quote comes from Douglas Miller, *op. cit.*, p. xxi.

8. The debate between Cuvier and Geoffroy, and the context in which the debate was set (including the meeting between Soret and Goethe) is described in *The Cuvier–Geoffroy Debate: French Biology in the Decades before Darwin* by Toby A. Appel (New York: Oxford University Press, 1987). It is a shame that this marvellous book is unread outside academia. Somebody should make it into a film.

9. Today, nobody believes that the skull has any particular relationship with vertebrae. The head seems to be a structure developmentally and morphologically unrelated to vertebrae. The braincase and parts of the base of the skull are made of cartilage separate from the cartilages that form the vertebrae: the jaws, face and much of the superficial bones of the skull are derived from neural crest cells.

10. Amazingly, recent work in comparative developmental genetics has suggested that Geoffroy's far-fetched ideas contain more than a grain of truth. The genes that in vertebrates direct the formation of dorsal structures, such as the spinal cord, have evolutionary correspondents in insects that organize the ventral ectoderm, and vice versa. This has

led some workers to posit a wholesale inversion of vertebrate structure (with respect to insect structure) very early in animal evolution. (See 'Inversion of dorsoventral axis?' by D. Arendt and K. Nübler-Jung, *Nature*, vol. 371, 1994, p. 26; and 'A common plan for dorsoventral patterning in Bilateria' by E. M. De Robertis and Y. Sasai, *Nature*, vol. 380, 1996, pp. 37–40.)

CHAPTER 5

1. Perhaps the most readable account of the life of Charles Darwin is *Darwin*, by Adrian Desmond and James Moore (London: Penguin, 1992). The history of evolution as well as genetics is told in breezy style by historian David L. Hull in *Science as a Process* (Chicago: University of Chicago Press, 1988).
2. The effort was vain: Fitzroy eventually killed himself.
3. I am indebted to Jack Cohen, who pointed out to me that profligate waste is the natural order of animal and plant reproduction: see his essay in *Nature*, vol. 411, 2001, p. 529.
4. Hull, *op. cit.*, p. 40.
5. This now cliché'd quote appears in a letter from Darwin to Joseph Hooker written in 1871, published in Francis Darwin (editor), *The Life and Letters of Charles Darwin*, vol. 3 (London: John Murray, 1888), p. 18.
6. Paracelsus, *Das Buch von der Gebärung der Empfindlichen Dinge in der Vernunft* (*c.*1520), vol. I, section i, pp. 261–2, cited on p. 59 of *Paracelsus: Essential Readings*, edited and translated by Nicholas Goodrick-Clarke (Berkeley, California: North Atlantic Books, 1999). I am grateful to Philip Ball for discovering this and other alchemical arcana.
7. Quoted on p. 131 of Elizabeth Gasking, *Investigations into Generation 1651–1828* (London: Hutchinson, 1967) from L. Spallanzani, *Tracts on the Natural History of Animals and Vegetables*, translated by J. Dalyell (Edinburgh, 1803).
8. F. J. Cole, *Early Theories of Sexual Generation* (Oxford: Clarendon Press, 1930), p. 176.
9. See my essay 'Aspirational thinking' (*Nature*, vol. 420, 2002, p. 611). The late Stephen Jay Gould was fond of exposing the conceits of progressive, destiny-driven evolution, as shown by his collection of man-from-the-apes commercials in *Wonderful Life* (New York: Norton, 1989).

10. I discuss this issue in depth on my book *In Search of Deep Time* (New York: Free Press, 1999), published in the UK as *Deep Time* (London: Fourth Estate, 2000).

11. *Form and Function: A Contribution to the History of Animal Morphology* by E. S. Russell (London: John Murray, 1916) is a typical example of a work written when disenchantment with Darwinism was at its deepest. Russell advocates a Lamarckian model of evolution and mentions with approval the work of Samuel Butler (1835–1902), the author of *Erewhon* and *The Way of All Flesh*, who was a prominent anti-Darwinian of the period. This is the same Butler whose views, along with those of Paley, were disparaged by Bateson in *Materials for the Study of Variation*. Despite Russell's views on evolution – which are, of course, nowadays seen as quite outmoded – *Form and Function* is a classic text on morphology, still cited today.

12. From the preface to W. Bateson, *Materials for the Studies of Variation treated with Especial Regard to Discontinuity in the Origin of Species* (London: Macmillan, 1894), p. v.

CHAPTER 6

1. See *Nature*, vol. 122, 1928, pp. 339–40; see also the anonymous obituary of Bateson in *Nature*, vol. 117, 1926, pp. 312–13.

2. See *Nature*, vol. 124, 1926, p. 171.

3. For more about the history of research into the origin of vertebrates, see my own book *Before The Backbone* (London: Chapman & Hall, 1996) The debate about vertebrate ancestry that took place at the Linnean Society appears in full in Gaskell, W. H. *et al.*, 'Discussion on the origin of vertebrates', *Proceedings of the Linnean Society of London*, session 122 (1909–1910), 1910, pp. 9–50. I am grateful to Stuart Baldwin for unearthing this gem for me.

4. Quoted from Gaskell *et al.*, *op. cit.*.

5. From the preface to W. Bateson, *Materials for the Studies of Variation treated with Especial Regard to Discontinuity in the Origin of Species* (London: Macmillan, 1894), p. vi.

6. *Ibid.*

7. *Ibid.*, p. 570.

8. *Ibid.*, p. 147.

9. *Ibid.*, p. 148.

10. *Ibid.*, p. 568 (emphasis in original).

11. *Ibid.*, p. 573.

12. Quoted on p. 53 of David L. Hull, *Science as a Process* (Chicago: University of Chicago Press, 1988).

CHAPTER 7

1. See *The Monk in the Garden: The Lost and Found Genius of Gregor Mendel, the Father of Genetics* by R. M. Henig (Boston: Houghton Mifflin, 2000).
2. Both J. B. S. Haldane (*Nature*, vol. 122, 1928, pp. 339–40) and Thomas Hunt Morgan (*Nature*, vol. 124, 1926, pp. 171–2) had views on what really happened when Bateson had his eureka moment on a train in 1900. Haldane implies that Bateson actually read Mendel's paper on the train, whereas Morgan suggests that Bateson had only read the paper by De Vries in which Mendel's work was mentioned. The account here – that Bateson had received De Vries's paper and then looked up Mendel – is the one given by David Hull in *Science as a Process* (Chicago: University of Chicago Press, 1988).
3. The life and works of Morgan are described in *Thomas Hunt Morgan, Pioneer of Genetics* by Ian Shine and Sylvia Wrobel (Lexington: The University Press of Kentucky, 1976).
4. Hence the popular scientific in-joke that time flies like an arrow, but fruit flies like a banana.
5. Shine and Wrobel, *op cit.*
6. It is possible that Mendel did not publish results that were less clear cut, but this may now be difficult to say given the destruction of many of his papers after his death.
7. W. S. Sutton, 'The chromosomes in heredity', *Biology Bulletin*, vol. 4, 1903, pp. 231–51.
8. There may even be more than one split point, but I confine my discussion to just one split point in each chromosome, to avoid it becoming more complicated than it needs to be.

CHAPTER 8

1. Myoglobin is the substance that carries oxygen within muscles, and is distinct from the haemoglobin that does the same job in the blood, although they are chemically related.
2. James D. Watson, *The Double Helix* (London: Penguin, 1999). This recent edition has an introduction by geneticist Steve Jones.
3. Walter Gratzer, *A Bedside Nature: Genius and Eccentricity in Science 1869–1953* (London: Macmillan Magazines, 1996).

4. J. D. Watson and F. H. C. Crick, 'A structure for deoxyribose nucleic acid', *Nature*, vol. 171, 1953, p. 737. Ever since this paper was published, whenever a *Nature* editor sees a line in a manuscript beginning 'It has not escaped our notice . . .' it is almost always remorselessly deleted.

5. An issue of *Nature* (vol. 421, 2003, pp. 396–453) commemorating the fiftieth anniversary of these events contains facsimiles of the Watson–Crick paper, as well as others which appeared at the time, and other material of interest exploring the scientific and cultural ramifications of the discovery of the structure of DNA.

6. See F. H. C. Crick, Leslie Barnett, S. Brenner and R. J. Watts-Tobin, 'General nature of the genetic code for proteins', *Nature*, vol. 192, 1961, pp. 1227–32.

7. Given the enormous differences between viruses and bacteria, it is exceedingly unfortunate and breathtakingly inappropriate that they tend to be grouped together (with any other microscopic organism) as *bugs*. The word 'bug' can be applied with justification to an insect of the order Hemiptera, a river in Eastern Europe, and, possibly, an unintended feature of a computer program. It should, however, not be applied to a microorganism, under any circumstances whatsoever, and anyone who uses the term in this sense is guilty of the kind of sloppy journalism that induces newsreaders (who are as routinely and shamefully ignorant of science as they are well-informed about politics) to use the terms 'virus' and 'bacterium' as if they were interchangeable – or worse, to use the word 'bacteria' (or, for that matter, 'criteria' or 'phenomena') as a singular.

CHAPTER 9

1. A. E. Garrod, *Inborn Errors of Metabolism* (London: Henry Froude and Hodder & Stoughton, 1923).

2. Beadle, Tatum, and Tatum's student Joshua Lederberg shared a Nobel prize in 1958 for their work.

3. This classic paper was published in the *Journal of Molecular Biology*, vol. 3, pp. 318–56, 1961.

4. The same lambda that hides out in the genome of strain K12λ of the bacterium *Escherichia coli*.

5. As everyone who does the laundry knows, there is no better way to get your washing *really* clean than to hang it outdoors in the sunshine. Sunshine is rich in UV and is a potent natural anti-bacterial agent.

CHAPTER 10

1. See the report by William Shawlot and Richard R. Behringer, 'Requirement for *Lim1* in head-organizer function', *Nature*, vol. 374, 1995, pp. 425–30.
2. For technical details on the homeo-box, see Guy Riddihough's article 'Homing in on the homeobox', *Nature*, vol. 357, 1992, pp. 643–4, and references therein.
3. Confusingly, some genes have both *paired*- and homeo-boxes, so they are simultaneously members of the *Hox* and *Pax* families.
4. Lewis's own account of his work can be found in 'A gene complex controlling segmentation in *Drosophila*', *Nature*, vol. 276, 1978, pp. 565–70.
5. Alexander Awgulewitsch and Donna Jacobs, '*Deformed* autoregulatory element from *Drosophila* functions in a conserved manner in transgenic mice', *Nature*, vol. 358, 1992, pp. 341–5; and Jarema Malicki, Luciano C. Cianetti, Cesare Peschle and William McGinnis, 'A human *HOX4B* regulatory element provides head-specific expression in *Drosophila* embryos', *Nature*, vol. 358, 1992, pp. 345–7.
6. J. Garcia-Fernández and P. W. H. Holland, 'Archetypal organization of the amphioxus *Hox* gene cluster', *Nature*, vol. 370, 1994, pp. 563–6.
7. A. Amores *et al.*, 'Zebrafish *hox* clusters and vertebrate genome evolution', *Science*, vol. 282, 1998, pp. 1711–14.
8. Sean B. Carroll, 'Homeotic genes and the evolution of arthropods and chordates', *Nature*, vol. 376, 1995, pp. 479–85.
9. The paper on gene regulation in fruit flies by Christiane Nüsslein-Volhard and Eric Wieschaus was published as 'Mutations affecting segment number and polarity in *Drosophila*' in *Nature*, vol. 287, 1980, pp. 795–801.
10. 20,000 was the total number of genes then thought to have been contained in the fruit fly's genome. Subsequent work has shown that the total is far fewer, around 13,600.
11. See 'Genetic control and evolution of sexually dimorphic characters in *Drosophila*' by Artyom Kopp, Ian Duncan and Sean B. Carroll, *Nature*, vol. 408, 2000, pp. 553–9.
12. The paper on the network properties of the segmentation module is 'The segment polarity network is a robust developmental module', by Georg von Dassow, Eli Meir, Edwin M. Munro and Garrett M. Odell, published in *Nature*, vol. 406, 2000, pp. 188–92.
13. My favourite expression of the democratic, even subversive, anonym-

ity of the internet – a consequence of the importance of the network over any of its components – is the well-known *New Yorker* cartoon in which two dogs sit in front of a monitor. 'On the internet,' says one, 'no one knows you're a dog.'

14. Actually, *Lim1*-mutant mice lack kidneys and gonads, but this is not apparent from a superficial examination.

15. This is, of course, an oversimplification. Many single-celled creatures develop inasmuch as changes in their environments allow them to select their fates, such as whether a bacterium should continue to feed or encapsulate itself into a quiescent cyst, or whether a bacteriophage should remain virulent or become an inactive passenger in the genome of its host.

CHAPTER 11

1. K. Padian, 'A daughter of the soil: Themes of deep time and evolution in Thomas Hardy's *Tess of the d'Urbervilles*', *Thomas Hardy Journal*, vol. 13, 1997, pp. 65–81.

2. See 'Whole-genome shotgun assembly and analysis of the genome of *Fugu rubripes*' by Samuel Aparicio and colleagues, *Science* vol. 297, 2002, pp. 1301–10; and the accompanying commentary, 'Vertebrate genomes compared' by S. Blair Hedges and Sudhir Kumar, *Science*, vol. 297, 2002, pp. 1283–5.

3. See E. G. Nisbet and N. H. Sleep, 'The habitat and nature of early life', *Nature*, vol. 409, 2001, pp. 1083–91.

4. To avoid cluttering up this chapter with footnotes to scientific reports on individual genomes, I refer the reader to the *Nature* online resource for genomics, http://www.nature.com/genomics/, as well as the online resource maintained by The Institute for Genomic Research (TIGR), http://www.tigr.org

5. See, for example, G. F. Joyce, 'Booting up Life', *Nature*, vol. 420, 2002, pp. 278–9; and S. A. Strobel 'Repopulating the RNA world', *Nature*, 2001, vol. 411, pp. 1003–6, and references therein.

6. Cairns-Smith's ideas are explored in his book *Seven Clues to the Origin of Life* (Cambridge: Cambridge University Press, 1990) and in a chapter, 'The origin of life: Clays', pp. 169–92, in *Frontiers of Life*, vol. 1, edited by D. Baltimore, R. Dulbecco, F. Jacob and R. Levi-Montalcini (New York: Academic Press, 2001). I am grateful to Professor Cairns-Smith for sending me a reprint of that chapter.

7. See R. D. Fleischmann and colleagues, 'Whole-genome random

sequencing and assembly of *Haemophilus influenzae* Rd.', *Science*, vol. 269, 1995, pp. 496–512.

8. I am not sure why, but I find the name *Reclinomonas americana* curiously attractive. It brings to mind 1950s kitsch and pop-art pieces such as the collage by Richard Hamilton entitled *Just What Is It That Makes Today's Homes So Different, So Appealing?*

9. See R. Ainscough and colleagues, 'Genome sequence of the nematode *Caenorhabditis elegans*: A platform for investigating biology', *Science*, vol. 282, 1998, pp. 2012–18.

10. This total was far less than the 20,000 assumed by Nüsslein-Volhard and Wieschaus in their mutational experiments.

11. See note 3 of Chapter 1.

12. See 'The draft genome of *Ciona intestinalis*: Insights into chordate and vertebrate origins' by Paramvir Dehal and colleagues, *Science*, vol. 298, 2002, pp. 2157–67; and 'Return of a little squirt', my commentary on this paper, in *Nature*, vol. 420, 2002, pp. 755–6.

13. See 'A new hominid from the Upper Miocene of Chad, central Africa' by Michel Brunet and colleagues, *Nature*, vol. 418, 2002, pp. 145–51.

14. See 'Molecular evolution of *FOXP2*, a gene involved in speech and language' by Wolfgang Enard and colleagues, *Nature*, vol. 418, 2002, pp. 869–72.

15. In his novel *The Fountains of Paradise*, Arthur C. Clarke jokes that religion is a by-product of the mammalian reproductive system.

16. See 'Growth processes in teeth distinguish modern humans from *Homo erectus* and earlier hominins' by Christopher Dean and colleagues, *Nature*, vol. 414, 2001, pp. 628–31.

17. See *The Wisdom of Bones* by Pat Shipman and Alan Walker (New York: Alfred A. Knopf, 1996) for a remarkable portrait of the life and times of *Homo erectus*.

CHAPTER 12

1. See 'Genetic control and evolution of sexually dimorphic characters in *Drosophila*' by Artyom Kopp, Ian Duncan and Sean B. Carroll, *Nature*, vol. 408, 2000, pp. 553–9.

CHAPTER 13

1. I have deliberately avoided making any connection between the evolution of sentience and the evolution of brains, because the evolution

of sentience is not necessarily restricted to the evolution of brains. One can imagine structured societies similar to those of social insects – ants, bees and wasps – in which individuals have very small brains but sentience might arise as a consequence of their collective actions. This point is elegantly made by Douglas Hofstadter in *Gödel, Escher, Bach: The Eternal Golden Braid* (London: Penguin, 1979).

2. See Sean Carroll's article 'Genetics and the making of *Homo sapiens*', *Nature*, vol. 422, 2003, pp. 849–57.

3. See Jared Diamond's article 'The double puzzle of diabetes', *Nature*, vol. 423, 2003, pp. 599–602.

4. See Gregory Stock's book *Redesigning Humans: Choosing our Children's Genes* (Boston: Houghton Mifflin, 2002).

5. In his article 'Artificial wombs: An out of body experience' (*Nature*, vol. 419, 2002, pp. 106–7), Jonathan Knight reports how technology to sustain life in early embryos is running in parallel with devices that can keep babies alive when born in an ever more premature condition. Perhaps, he wonders, the two efforts will meet in the middle and we shall have artificial wombs in which gestation can proceed entirely outside the human body. If this happens, human beings in their mode of reproduction will be no different from a chicken or any other animal that lays external eggs – a curious vindication of Harvey's dictum.

6. The social consequences of elective genetic modification have provided a major theme in contemporary science fiction. As a genre, science fiction does not predict the future: rather, it provides the expression, in the context of the future, of current preoccupations about the impact of technology on society. Immediately after World War II, for example, SF was full of post-apocalyptic stories of worlds recovering from nuclear calamity, whereas today the genre is preoccupied with what people in the field call 'posthumanity'. To celebrate the millennium, I ran a column in *Nature* called 'Futures' in which writers, almost all professional SF authors, were invited to explore, in the form of a short, fictional vignette, how technology might affect the way we live in the next millennium. 'Futures' ran from November 1999 to December 2000, and of the fifty-eight stories published, eleven explored the consequences of improvements in the technology of reproduction, including the modification of the human form, on ethics, politics, religion, personality and society. Only three stories involved direct contact with aliens – usually regarded as the SF staple – and one of those used aliens as a vehicle for comedy.

Index